Essential Maths Skills
for AS/A-level
Geography

Helen Harris

PHILIP ALLAN FOR
HODDER
EDUCATION
AN HACHETTE UK COMPANY

Acknowledgements

p.36 Figure 3.1 © FAO 2006 'The world's forests', http://www.fao.org/forestry/fra/41256/en/, accessed 11 July 2016; p.54 Figure 4.10 contains public sector information licensed under the Open Government Licence v1.0 http://www.ons.gov.uk/ons/rel/regional-trends/regional-economic-analysis/the-spatial-distribution-of-industries/sty-employment-by-industry.html; p.76 Figure 6.1 contains public sector information licensed under the Open Government Licence v1.0. http://www.metoffice.gov.uk/media/pdf/n/9/Fact_sheet_No._14.pdf, page 17.

Philip Allan, an imprint of Hodder Education, an Hachette UK company, Blenheim Court, George Street, Banbury, Oxfordshire OX16 5BH

Orders

Bookpoint Ltd, 130 Park Drive, Milton Park, Abingdon, Oxfordshire OX14 4SE
tel: 01235 827827
fax: 01235 400401
e-mail: education@bookpoint.co.uk

Lines are open 9.00 a.m.–5.00 p.m., Monday to Saturday, with a 24-hour message answering service. You can also order through the Hodder Education website: www.hoddereducation.co.uk

ISBN 978-1-4718-6355-4

First printed 2016
Impression number 5 4 3
Year 2020 2019 2018 2017

Typeset in India

Cover illustration: Barking Dog Art

Printed in India

Hachette UK's policy is to use papers that are natural, renewable and recyclable products and made from wood grown in sustainable forests. The logging and manufacturing processes are expected to conform to the environmental regulations of the country of origin.

Contents

Exam-style questions

Appendix 1

Appendix 2

Appendix 3

Appendix 4

Introduction

The analysis of data is fundamental to the study of geography, as it can support the development of geographical concepts and theories. In fieldwork, mathematical techniques help you to analyse the data you have collected so that valid conclusions can be drawn. Being able to use mathematical methods accurately and confidently will allow you to be more precise in your geographical analysis. This book will help you to become familiar with the mathematical concepts that are fundamental to AS and A-level geography courses, and will provide you with opportunities to practise the quantitative skills required for examinations, practical fieldwork and the A-level independent investigation in geography. It covers descriptive tools for measuring location, dispersion and concentration, as well as statistical tests which are commonly used to investigate the relationships and differences between data sets. The use of mathematics ranges across your studies of human, physical and integrated geography.

This book does not assume any prior knowledge of the mathematical concepts and techniques, or of the geographical theory in which they are set. Each chapter will introduce a particular concept, provide step-by-step worked examples and then offer opportunities to practise applying the techniques through guided questions and practice questions of increasing difficulty.

You can use this book in various ways. An initial read-through of the concepts covered will help you familiarise yourself with the maths required in AS and A-level geography. Throughout your course this book can be used alongside your textbook and other support materials for additional maths practice. The guided and practice questions within the chapters and the exam-style questions at the end provide a specialised revision resource. All mathematical concepts are embedded in relevant geographical contexts, which will build on your knowledge and understanding of the subject. A list of key terms is provided in Appendix 3.

Practical fieldwork and an independent investigation (sometimes called the 'non-examined assessment' or 'A-level research project') feature in all geography A-level specifications. This book will help you to select and apply appropriate statistical tests to validate your findings in both of these elements (see Appendix 2). For each technique, the book gives a clear step-by-step guide which will enable those unfamiliar with the use of maths in geography to become confident and competent in data analysis.

The content of the book is suitable for all exam boards, including AQA, Edexcel, Eduqas and OCR. The geographical contexts draw on core and optional topics and the grid in Appendix 4 will help you identify the mathematical skills relevant to your particular board.

To get into good habits for the exam, be sure to always give units (such as % or $) and show working with your answers. The book also emphasises that, in geography, a logical next step in the analysis is to seek reasons for the patterns and relationships identified.

Full worked solutions to the guided and practice questions and the exam-style questions can be found online at www.hoddereducation.co.uk/essentialmathsanswers.

1 Understanding data

Nominal, ordinal and interval data

Data is central to the study of geography. When geographical research or fieldwork is undertaken, data is collected. In data analysis, **variables** are the quantities that have been measured — for example, the numbers of various plant species or the sizes of stones in a river. For each variable, a score or value is recorded from each item sampled (such as each plant or each stone) and the collection of scores or values constitutes the **data set**. This set of data will then be subjected to further analysis. There are many statistical measures and tests which can help us interpret the data so that conclusions can be drawn. The 'type' of data, as well as the sample size (the size of the data set), can influence the choice of the most appropriate statistical test to use.

Data can be classified into the following types:

- nominal
- ordinal
- interval
- ratio

Nominal data

This is the most basic kind of data. With nominal data, the 'values' that variables can take belong to separate categories which are given labels or names with no particular sense of order. For example, the values can be words describing types of rock or types of land use. Sometimes the categories are labelled with numerical codes, such as '1' for industrial land, '2' for retail space and so on; however, these labels, despite being numbers, are not actually quantitative and cannot be meaningfully ordered relative to one another.

Frequencies or percentages are often used to summarise nominal data, for example, in statements like '50% of the land use is retail'.

Ordinal data

With ordinal data, the values of variables can be put into an order. The numbers on an ordinal scale describe a hierarchy, such as from low to high. The intervals or gaps between consecutive values on the scale need not be the same, which is the case for the river flow rates shown in Table 1.1. Station 1 recorded the highest flow rate (the shortest time needed to cover 10 metres) and station 5 the lowest flow rate, so the ordinal data values in the right column are flow rates ordered from high to low.

Table 1.1 River flow data at Skirden Beck

Station	Time (seconds) to cover 10 metres
1	12
2	15
3	19
4	22
5	25

Ordinal data values do not even have to be numbers. For example, in the parking survey of visitors to a tourist town shown in Table 1.2, the values of the 'Quality of parking' variable are qualitative. In this case you cannot calculate the mean.

Table 1.2 Parking survey in Clitheroe

Quality of parking	Number of people
Poor	3
Average	5
Good	12

In both examples, it is the fact that the values can be ranked in **order** which makes them ordinal data.

Interval data

Like ordinal data, interval data can be ordered, but in this case the interval between consecutive values on the measurement scale is constant. For example, if you record times from a clock, temperatures from a thermometer or pH values of water samples, the resulting data is interval data.

Ratio data

Ratio data is a particular type of interval data which is measured on a scale that allows numbers to be compared with each other using ratios. For example, temperatures measured on the Celsius scale do **not** form a ratio data set, because 20°C is not 'twice as hot' as 10°C; however, masses measured in kilograms do form a ratio data set, since 80 kg is twice as heavy as 40 kg. Similarly, if you count the number of houses of each type on a housing estate, you get a set of ratio data, because you can make statements such as 'there are twice as many detached houses as terraced houses'.

Figure 1.1 summarises the main characteristics of each type of data and the statistical terms that are often used to describe them.

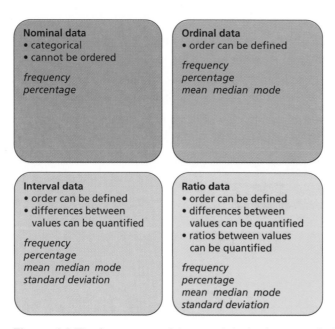

Figure 1.1 The four types of data and their characteristics

Ratios and fractions

Ratios

A ratio is a comparison of two numbers of the same kind and states how much of one thing there is relative to another thing. For example, in a farm with 12 fields, if there are ten fields of wheat and two fields of barley, then the ratio of wheat fields to barley fields is 10 to 2. This is usually written as $10:2$, with a colon separating the numbers being compared.

In this example, the ratio can be simplified — both numbers are divisible by the same whole number, 2. Therefore, after performing the divisions by 2, the ratio can be expressed in the simpler form $5:1$.

It is usual to show ratios in their lowest (that is, most simplified) form. For example, for the ratio $100:25$ we would write $4:1$ and for the ratio $9:27$ we would write $1:3$.

If the ratio cannot be reduced to a lower form, then just leave in its original form, such as $25:7$ or $39:2$ — in both cases the two numbers in the ratio cannot be divided by the same whole number.

A dependency ratio is an application of ratios specific to population geography. A dependency ratio is represented by the equation:

$$\text{dependency ratio} = \frac{\text{non-working population (under 15 and over 65)}}{\text{working population (15–65)}} \times 100$$

The result measures the extent to which the non-working population is economically dependent on the working population and is expressed as the number of dependants **per 100** working people. Developed countries usually have a ratio of around 50 e.g. UK 54, while developing countries can have a ratio as high as 80+ e.g. Sudan 85.

Fractions

A fraction is a **proportion** or part of a whole. It is a numerical value that is not a whole number.

In the farm example above, only a proportion of the fields are used to grow wheat: 10 of the 12 parts. This is written as the fraction $\frac{10}{12}$.

Similarly to simplifying ratios, we can simplify fractions by dividing both numbers by the same whole number (if possible). So the fraction $\frac{10}{12}$ can be simplified to $\frac{5}{6}$.

Worked example

In a survey of shoppers in Clitheroe, 30 people were asked if they lived in Clitheroe or outside the town. The results were:

- **20 people lived in Clitheroe.**
- **10 people lived outside Clitheroe.**

What is the ratio of respondents who live in Clitheroe to the total number of respondents?
What fraction of respondents live in Clitheroe?

respondents who live in Clitheroe : total respondents

$= 20 : 30$

$= 2 : 3$ (after dividing both numbers by 10)

Expressed as a fraction, $\frac{2}{3}$ of the respondents live in Clitheroe.

B Practice question

1 A group of geography students conducted a housing survey as part of an investigation into the characteristics and profile of a local commuter town. Estate agents provided information on the types and cost of housing for sale in the area, shown in Table 1.3, where D = detached, SD = semi-detached, T = terraced and F = flat.

Table 1.3 Types and cost of housing for sale

Type of property	Number of bedrooms	Cost (£)
D	5	750 000
D	3	750 000
D	4	680 000
D	4	500 000
SD	3	380 000
SD	4	350 000
SD	3	315 000
SD	4	285 000
T	3	230 000
SD	3	219 000
T	2	190 000
F	3	170 000

a What is the ratio of semi-detached to all properties?
b What is the ratio of 4-bedroom houses to all properties? Simplify the ratio and express it as a fraction.
c What is the ratio of semi-detached to detached houses?
d Expressed as a fraction, how many of the properties in the survey are terraced?

Percentages

A percentage (%) is a proportion or fraction that is expressed as a 'number of parts out of 100'. So 1% means 1 part per 100, 50% is 50 parts per 100 and 25% is 25 parts per 100.

Another way to express a number of parts out of 100 is as a decimal, so a percentage can also be written as a decimal number.

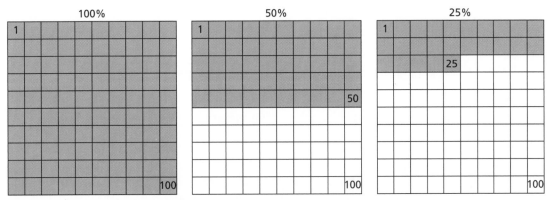

Figure 1.2 Box diagrams showing percentages of 100%, 50% and 25%

Uses of percentages in geography

Percentages are frequently used in geography to:

■ enable comparisons to be made between data sets with different totals by converting raw figures to percentages
■ enable comparisons to be made between data sets with different totals by converting percentages to raw figures
■ calculate the amount of change over time in a data set

Converting between percentages and decimals

To change a percentage to a decimal, **divide** the number in front of the % sign by 100.

To change a decimal to a percentage, **multiply** the decimal by 100 and add a % sign.

A Worked examples

a **Write 32% as a decimal.**
To change a percentage to a decimal, **divide** by 100.

32% as a decimal is
$32 \div 100 = 0.32$

b **Write 0.74 as a percentage.**
To change a decimal to a percentage, **multiply** by 100.

0.74 as a percentage is
$0.74 \times 100\% = 74\%$

Practice questions

1 Express each of the following percentages as a decimal.

   a 16%   b 4%

2 Express each of the following decimals as a percentage.

   a 0.14   b 0.02

Converting raw figures to percentages

The data in Table 1.4 was collected as part of a study to investigate the extent to which a local place had been affected by counter-urbanisation.

Table 1.4 Traffic count data from four collection points at 8.30 a.m.

Data collection point	Car/ motorcycle	Service bus	School bus, coach or mini-bus	Heavy goods vehicle (lorry, van, tractor)	Total
1	125	7	3	13	148
2	132	5	1	16	154
3	185	11	9	10	215
4	162	1	5	10	178

As the data collection points all have different totals, in order to compare the traffic count at the four locations it helps to convert the figures into percentages.

A Worked example

From the data in Table 1.4, what percentage of vehicles counted at data collection point 1 are cars/motorcycles?

Step 1: identify the relevant figures.

The number of cars/motorcycles at data collection point 1 is 125.
The total number of vehicles counted at data collection point 1 is 148.

Step 2: divide the number of cars/motorcycles by the total number of vehicles.

   $125 \div 148 = 0.84$

Step 3: multiply this decimal by 100 and add a % sign.

   $0.84 = (0.84 \times 100)\% = 84\%$

So 84% of vehicles counted at data collection point 1 are cars/motorcycles.

B Guided question

Copy out the workings and complete the answers on a separate piece of paper.

1 **From the data in Table 1.4, find the percentages of vehicles counted at data collection points 2, 3 and 4 that are cars/motorcycles. Which collection point had the highest percentage of cars and motorcycles?**

For collection point 2:

Step 1: the number of cars/motorcycles is 132. The total number of vehicles is _____

Step 2: 132 ÷ _____ = _____

Step 3: (_____ × 100)% = _____%

For collection point 3:

Step 1: the number of cars/motorcycles is _____. The total number of vehicles is 215.

Step 2: _____ ÷ 215 = _____

Step 3: (_____ × 100)% = _____%

For collection point 4:

Step 1: the number of cars/motorcycles is _____. The total number of vehicles is _____

Step 2: _____ ÷ _____ = _____

Step 3: (_____ × 100)% = _____%

Comparing the results above, collection point _____ had the highest percentage of cars and motorcycles.

Converting percentages to raw figures

In some situations, equal percentages can mean very different actual numbers. For example, in Figure 1.3 on page 14, the same proportion (25%) of the populations of China, the UK and Kuwait represent vastly different numbers of people, because the total populations of the countries are very different.

Consider the following statements:

- In 1993 the amount of fossil fuel the world used was 8124 MTOE; by 2011 it had increased by 42%.
- In 1993 the amount of renewable energy the world used was 991 MTOE; by 2011 it had increased by 56%.

Here MTOE stands for 'million tonnes of oil equivalent'. This is a standard measure of energy sources which allows comparisons to be made.

The statements give the percentage increases in amounts of fossil fuel and renewable energy, but what is the actual increase in size in MTOE? To answer this question we need to convert the given percentages into raw figures.

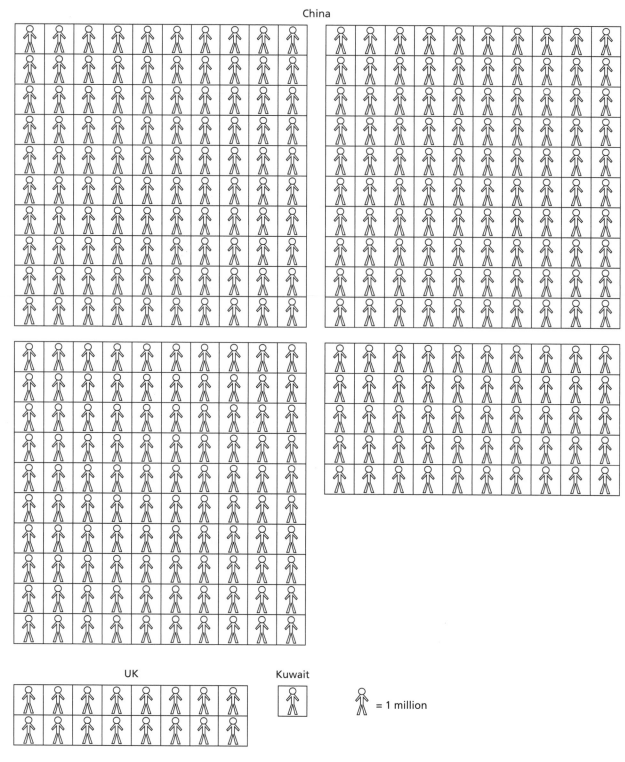

Figure 1.3 25% of the populations of China, the UK and Kuwait

Worked example

Using the information on page 13, find the increase in the amount of fossil fuel used from 1993 to 2011 in MTOE.

Step 1: convert the percentage to a decimal, i.e. divide by 100.

$42 \div 100 = 0.42$

Step 2: multiply this decimal by the initial amount of fossil fuel in 1993.

$0.42 \times 8124 = 3412$

Therefore, between 1993 and 2011, the global use of fossil fuel increased by 3412 MTOE.

(Don't forget to include the units of measurement in your answer.)

 Guided question

Copy out the workings and complete the answers on a separate piece of paper.

1 **Based on the second statement above, what is the increase in consumption of renewable energy between 1993 and 2011 in MTOE?**

Step 1: convert the percentage to a decimal, i.e. divide by 100.

_____ $\div 100 =$ _____

Step 2: multiply this decimal by the initial amount of renewable energy used in 1993.

_____ $\times 991 =$ _____

Between 1993 and 2011, the global consumption of renewable energy increased by _____ MTOE.

Calculating percentage changes

In the previous Worked example and Guided question, you calculated the actual amount of change over time from a given percentage change. In this section we do the reverse and calculate percentage changes from raw figures collected over time.

Table 1.5 shows the number of tourist arrivals to different regions of the world in 2010 and 2014. In all regions there has been an increase in the number of tourist arrivals between 2010 and 2014, but which regions have experienced the highest (and lowest) levels of growth in tourism? To measure 'growth' over time, and especially to compare growth rates, it is common to use percentage changes.

Table 1.5 International tourist arrivals (millions) to different world regions in 2014

Region	2010	2014	% change
Europe	497	584	
Asia and the Pacific	206	263	
Americas	156	182	
Africa	28	56	
Middle East	34	50	+47%
Total	921	1 135	

Source: World Tourism Organization (UNWTO)

Worked example

From the data in Table 1.5, calculate the percentage change in tourist arrivals to Europe between 2010 and 2014.

The percentage change over time is calculated in the following steps.

Step 1: identify the appropriate starting value.

In this example it is the tourist arrivals to Europe in 2010: 497 (million)

Step 2: calculate the difference between the starting value and the value at the comparison time. In this case the comparison time is the year 2014, in which the value for Europe is 584, so calculate

$584 - 497 = 87$

Step 3: divide the difference by the starting value.

$87 \div 497 = 0.18$

Step 4: multiply by 100 to get the percentage change from the starting time to the comparison time.

$0.18 \times 100 = 18$

So between 2010 and 2014 there has been an 18% increase in the number of tourists arriving in Europe. This can be entered in the final column of Table 1.5 as '+18%' for Europe.

Note that because the difference calculated in Step 2 is positive, the result is a percentage increase.

B **Guided question**

Copy out the workings and complete the answers on a separate piece of paper.

1 **From the data in Table 1.5, calculate the percentage change in tourist arrivals to Asia and the Pacific between 2010 and 2014.**

Step 1: identify the appropriate starting value.

The number of tourist arrivals to Asia and the Pacific in 2010 is _____ million.

Step 2: calculate the difference between the starting value and the value at the comparison time.

The comparison time is the year 2014, in which the value for Asia and the Pacific is _____, so calculate _____ − _____ = _____

Step 3: divide the difference by the starting value.

_____ ÷ _____ = _____

Step 4: multiply by 100 to get the percentage change from the starting time to the comparison time.

_____ × 100 = _____

So between 2010 and 2014 there has been a _____ % increase/reduction in the number of tourists to Asia and the Pacific.

C Practice questions

2 In Table 1.4, showing traffic count data, collection point 3 had the highest total count of traffic. Express this as a percentage of the total traffic count obtained from all four data collection points.

3 a Using the data in Table 1.5 on international tourist arrivals, calculate the percentage change between 2010 and 2014 in tourist arrivals to
   i   the Americas
   ii Africa

   b Which region had the highest percentage change in international tourist arrivals between 2010 and 2014?

4 Table 1.6 shows the carbon dioxide emissions of selected countries.

Table 1.6 CO_2 emissions (in millions of tonnes) in 1981 and 2011

Country	1981	2011
Brazil	172	439
China	1452	9020
France	455	389
India	375	2074

Source: World Bank

Calculate the percentage change in CO_2 emissions between 1981 and 2011 for:

a Brazil   c France
b China   d India

For each country, express your answer as a concluding statement of the form:

'There has been a _____% increase/decrease in CO_2 emissions in _____ between 1981 and 2011.'

Densities

Density is understood and calculated in different ways for different subjects. Geographers are primarily interested in how densely items (such as people, plants or other objects) are distributed **spatially**, so density is usually taken to mean the number of items **per unit area**. This measure is frequently used in geography to analyse spatial patterns and concentrations. We can then try to find out why a pattern has occurred.

To calculate spatial density, divide the number of items by the area. The result is expressed as a number per unit area, such as per square kilometre (km^2).

Density calculations are frequently used in geography to:

■ compare the population densities of different countries, usually on the basis of people per km^2. The calculation for this is:

   number of people ÷ the area they occupy = population density

- define urban and rural areas. The threshold used can vary widely across countries: in China an urban area is defined as an area with a population density of 1500 people per km² or higher, whereas in Canada the density threshold for an urban area is 400 people per km².
- compare the housing stock in different areas to identify possible situations of overcrowding.

 Worked example

Singapore has a population of 5 469 700 and a land area of 707 km². Iceland has a population of 327 589 and a land area of 100 250 km². Calculate the population densities of Singapore and Iceland.

Singapore:

5 469 700 ÷ 707 = 7736 people per km²

Iceland:

327 589 ÷ 100 250 = 3 people per km²

B Practice questions

1 a Using the figures in Table 1.7, calculate the population densities of the three English districts.

Table 1.7 Population and land area of three English districts

District	Population	Land area (km²)
Harrogate	157 900	1 316
Cornwall	532 300	3 563
South Tyneside	148 100	64

Source: Office for National Statistics (ONS)

b Eurostat defines an urban area as one with a population density in excess of 150 people per km². Using this threshold, determine which of the districts are rural and which are urban.

2 Using the figures in Table 1.8, calculate the housing density, in number of dwellings per hectare (ha), for the three London boroughs.

Table 1.8 Housing in three London boroughs

Borough	Number of dwellings	Land area (ha)
Camden	101 210	2 179
Brent	114 040	4 323
Bromley	136 450	15 013

Source: London DataStore

3   Using the data in Table 1.9, calculate the agricultural density (number of people per km² of agricultural land) for the three countries.

Table 1.9 Population and area of agricultural land for selected countries

Country	Population	Agricultural land (km²)
Australia	23 000 000	460 938
Kenya	44 000 000	56 914
India	1 252 000 000	1 486 595

Source: World Bank

Geographical application

Fieldwork and the independent investigation

Ratios, fractions, percentages and densities provide straightforward ways of summarising data. By making such **descriptive** summaries, geographers can comment on the general character of a data set and take preliminary steps in choosing appropriate statistical techniques to perform further analysis, comparisons and hypothesis testing.

2 Measures of central tendency

Mean, median and mode

Measures of **central tendency** are used in geography to identify simple features of data or differences between data sets. These measures give a single figure which provides a useful **summary** of the data set. When tables of data are generated from fieldwork or collected from secondary sources, a preliminary overview can be obtained by calculating measures of central tendency. For example, if you wanted to study differences in stream flow over a number of sites or variations in property values in different parts of an urban area, it would be convenient to have a single number representing all the data collected from each location.

Geographers mostly use the following three standard measures of central tendency:

- **Mean** — the most common measure, calculated by adding up all the individual data values and dividing by the total number of data items. The mean is what is usually meant by the term 'average'.
- **Median** — the middle value in the data set. To find the median, arrange all the data items in order (a 'rank order') and identify the middle value in the sequence. If there is an even number of data items, the median is halfway between the two middle values; that is, add them together and divide by 2.
- **Mode** — the most straightforward measure, obtained by identifying the most frequently occurring value in a data set or, if the data has been put into categories, the most common category.

A Worked examples

Table 2.1 contains secondary data on the number of migrants entering the UK between 2005 and 2014. You want to use some basic measures of central tendency to obtain an overall summary of the data set.

Table 2.1 Immigration to the UK, 2005–2014

Year	Number of immigrants to the UK (thousands)
2005	567
2006	596
2007	574
2008	590
2009	567
2010	591
2011	566
2012	498
2013	526
2014	632

Source: Office for National Statistics (ONS)

Full worked solutions at www.hoddereducation.co.uk/essentialmathsanswers

a **Calculate the mean value of this data set.**

Step 1: add the numbers of immigrants in all the years to find the total number for the 10-year period. Remember that the data in the table is in thousands.

$567 + 596 + 574 + 590 + 567 + 591 + 566 + 498 + 526 + 632 = 5707$ (thousand)

Step 2: divide the total by the number of years, in this case 10, to find the mean.

$5707 \div 10 = 570.7$ (thousand)

Rounding to the nearest whole number, we find that the mean annual number of immigrants to the UK from 2005 to 2014 is 571 thousand.

b **Calculate the median value of this data set.**

Step 1: order the data values from lowest to highest.

498, 526, 566, 567, 567, 574, 590, 591, 596, 632

Step 2: the median is the middle value in this ordered list.
Because there is an even number of values (10), the median is halfway between the 5th and 6th values.

The 5th value is 567; the 6th value is 574. Add them and divide by 2:

$567 + 574 = 1141$

$1141 \div 2 = 570.5$

Rounding to the nearest whole number, we find that the median annual number of immigrants to the UK from 2005 to 2014 is 571 thousand.

c **Find the mode of this data set.**

From the ordered list in Worked example **b**, it can be seen that 567 is the only repeated value (occurring twice), so the mode is 567 thousand.

Comparing the mean, median and mode

Mean

- Calculation of this average makes use of all the data values and gives you a simple overview of the whole data set.

- It works best when the data values span a fairly narrow range. If there are several exceptionally high or low values (called 'anomalies' or 'outliers'), the mean will not be very representative of the data set.

Median

- As the median is the middle value when the data items are ranked from lowest to highest, it is quite straightforward to calculate.

- It is not affected by extreme values in the data set.

Mode

- The mode is the easiest measure of central tendency to obtain and it can be used for categorical data, whereas the mean and median can only be used with ordinal (including interval and ratio) data.

- It is not affected by extreme values in the data set.

- However, a data set can have no modal value (if each data value occurs only once) or more than one modal value, which could be confusing.

As a geographer, after calculating a measure or measures of central tendency, you should then use this information to analyse the data further. For instance, in the Worked example, you could examine the data to see if certain years had well above (632 thousand in 2014) or well below (498 thousand in 2012) the average intake of migrants and the next step would be to investigate the possible reasons for this.

B Guided question

Copy out the workings and complete the answers on a separate piece of paper.

1   **Table 2.2 shows rainfall totals recorded in Oban and Norwich for each month in 2014.**

Table 2.2 Monthly rainfall in Oban and Norwich in 2014

Month	Rainfall in Oban (mm)	Rainfall in Norwich (mm)
January	150	42
February	75	39
March	98	44
April	68	39
May	76	36
June	89	52
July	55	45
August	80	51
September	98	63
October	130	59
November	128	65
December	110	61

a   **Calculate the mean monthly rainfall for Oban and for Norwich.**

For Oban:

Step 1: add up the rainfall amounts in each month.

$150 + 75 + 98 + 68 + 76 + 89 + 55 + 80 + 98 + 130 + 128 + 110 = $ _____ mm

Step 2: divide the total by the number of months.

_____ ÷ 12 = _____

The mean monthly rainfall for Oban is _____ mm.

For Norwich:

Step 1: add up the rainfall amounts in each month.

$42 + $__$ + $__$ + $__$ + $__$ + $__$ + $__$ + $__$ + $__$ + $__$ + $__$ + 61 = $ _____ mm

Step 2: divide the total by the number of months.

_____ ÷ 12 = _____

The mean monthly rainfall for Norwich is _____ mm.

b **Calculate the median monthly rainfall for Oban and for Norwich.**

For Oban:

Step 1: order the values from lowest to highest.

55, 68, 75, 76, 80, 89, 98, 98, 110, 128, 130, 150

Step 2: because there is an even number of values (12), the median is halfway between the 6th and 7th values (so that there are five values above and five values below).

The 6th value is ____; the 7th value is ____. Add them and divide by 2:

____ + ____ = ____

____ ÷ 2 = ____

The median monthly rainfall for Oban is _____mm.

For Norwich:

Step 1: order the values from lowest to highest.

36, ____, ____, ____, ____, ____, ____, ____, ____, ____, ____, 65

Step 2: because there is an even number of values (12), the median is halfway between the 6th and 7th values.

The 6th value is ____; the 7th value is ____. Add them and divide by 2:

____ + ____ = ____

____ ÷ 2 = ____

The median monthly rainfall for Norwich is _____mm.

c **Identify the modal monthly rainfall for Oban and for Norwich.**

For Oban:

Step 1: order the values from lowest to highest.

55, 68, 75, 76, 80, 89, 98, 98, 110, 128, 130, 150

Step 2: identify the most common value in the data set.

The value ____ occurs twice; all the other values occur only once.

The modal monthly rainfall for Oban is _____mm.

For Norwich:

Step 1: order the values from lowest to highest.

36, ____, ____, ____, ____, ____, ____, ____, ____, ____, ____, 65

Step 2: identify the most common value in the data set.

The modal monthly rainfall for Norwich is _____mm.

d **What is the difference in average annual rainfall between Oban and Norwich?**

If it is not specified which kind of 'average' is meant, usually we take it to be the mean. In this case, let us calculate the difference between the means, the medians and the modes for comparison.

Difference in mean rainfall of Oban and Norwich = ____ − ____ = ____ mm

Difference in median rainfall of Oban and Norwich = ____ − ____ = ____ mm

Difference in modal rainfall of Oban and Norwich = ____ − ____ = ____ mm

C Practice questions

2   Table 2.3 shows data for the top ten countries in imports and exports in 2013.

Table 2.3 Top ten exporters and importers in 2013 (values are in billions of US dollars)

Rank	Exporters	Value	Rank	Importers	Value
1	China	2 209	1	United States	2 329
2	United States	1 580	2	China	1 950
3	Germany	1 453	3	Germany	1 189
4	Japan	715	4	Japan	833
5	Netherlands	672	5	France	681
6	France	580	6	UK	655
7	Republic of Korea	560	7	Hong Kong	622
8	UK	542	8	Netherlands	590
9	Hong Kong	536	9	Republic of Korea	576
10	Russian Federation	523	10	Italy	477

Source: World Trade Organization (WTO)

a   What is the **mean** value of the
  i   exports?
  ii imports?

b   What is the **median** value of the
  i   exports?
  ii imports?

c   Do the median values for imports and exports accurately reflect the range of data in the table?

3   Table 2.4 shows fieldwork data collected from three different sites on a beach in North Wales. The values in the second, third and fourth columns are the numbers of pebbles of each size found at that particular site.

Table 2.4 Data from a survey of pebble sizes on a beach in North Wales

Long axis (mm)	Site 1	Site 2	Site 3
<10	67	17	6
10–19	54	24	12
20–39	20	23	47
>40	14	23	53
	Sea	$\longrightarrow$	**Inland**

What is the modal category for each of the three sites?

TIP

In mathematical notation the mean is written as

$$\overline{x} = \frac{\Sigma x}{n}$$

where n is the number of data items and Σ (Sigma, the Greek capital letter 'S') stands for 'sum', so that Σx represents the sum of all the data values 'x'.

Frequency distributions

Frequency refers to how often something happens. In terms of data description, the frequency is the number of times a particular data value occurs in a data set.

The values in a data set, along with their corresponding frequencies, can be presented in a table or a chart.

Frequency tables

If there are not many distinct values in a data set, then each value can be displayed next to its frequency in a simple frequency table. However, most data sets generated from fieldwork will contain a large number of individual values, so for convenience we first group the values into non-overlapping **classes**. Then we present the classes, together with the frequency of each class, in a **grouped frequency table** or **frequency distribution table**. Before making a grouped frequency table, we need to decide on:

- the number of classes needed for the data set
- the width of (i.e. the interval of values spanned by) each class

If we use too few classes, we may lose a lot of the detail in the data set. On the other hand, dividing the data into too many classes may obscure the overall pattern of distribution.

Here are two simple calculations to help you find a suitable number of classes and the width of those classes:

- number of classes = square root of the total number of data values ($\sqrt{n}$)
- class width = range of the data values (difference between the highest and lowest values) divided by the number of classes

---

Ⓐ Worked example

A group of students set out to examine the sorting of beach material across a beach profile in North Wales. A random sample of 50 pebbles was collected and measured. The raw data (diameter of each pebble in centimetres rounded to nearest whole number) is shown in Table 2.5. Organise the data in a frequency distribution table.

Table 2.5 Random sample of pebbles on Borth Beach

12	5	10	10	10	5	5	1	12	6
8	10	6	12	12	6	2	7	7	3
8	8	2	5	7	7	14	5	3	3
9	8	5	7	2	6	5	5	10	6
7	9	7	9	12	6	11	4	6	14

Step 1: decide on the number of classes.

$$\text{Number of classes} = \sqrt{\text{total number of data values}}$$
$$= \sqrt{50} = 7.07 \approx 7$$

Step 2: determine the width of each class.

$$\text{Range of data values} = \text{highest value} - \text{lowest value} = 14 - 1 = 13$$

$$\text{Class width} = \text{range of data values} \div \text{number of classes}$$
$$= 13 \div 7 = 1.86 \approx 2$$

So group the data into 7 classes, in intervals of 2.

Step 3: enter the classes and the number of values belonging to each class (the frequency) in a table.

For example, 4 of the 50 values are in class '1–2' (the values 1, 2, 2, 2).

Table 2.6 Frequency table for pebble sample from Borth Beach

Class (cm)	Frequency
1–2	4
3–4	4
5–6	15
7–8	11
9–10	8
11–12	6
13–14	2

In geography, frequency tables are used to organise large data sets and identify the general shape of the data distribution. Measures of central tendency such as the median and mode can easily be obtained from a frequency table. Frequency tables are also used to prepare data for presentation in a **histogram** or **choropleth map**, allowing further analysis of patterns and trends.

Histograms

A frequency distribution table can provide a lot of useful information on its own; for example, from Table 2.6 it can be seen that 5–6 cm is the modal class of the pebble diameters and that the median pebble size (between the 25th and 26th values) lies in the 7–8 cm class. Plotting the frequency information in a chart would, however, give a clearer visual impression of the distribution. The type of chart normally used to display frequency data is a **histogram**.

To make a histogram, mark the classes of data values along the horizontal axis (*x*-axis) and mark the frequencies along the vertical axis (*y*-axis). For each class, draw a rectangle (bar) based on the *x*-axis, with width equal to the class interval and height equal to the frequency of that class. The histogram for the pebble data in Table 2.6 is shown in Figure 2.1.

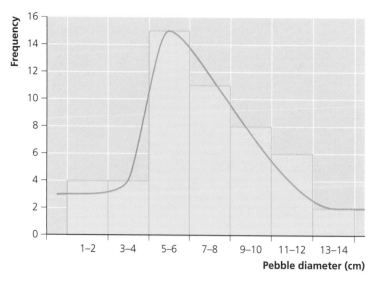

Figure 2.1 Histogram of the pebble data with a frequency distribution curve

Frequency distribution curves

After a histogram has been drawn, a **frequency distribution curve** can be added, as shown in Figure 2.1, to further clarify the shape of the distribution. There are three main types of frequency distribution curve.

Normal distribution

If the curve is **symmetrical** and 'bell-shaped', it is described as a **normal distribution**. Normal distribution curves can vary as to how tall or flat and how narrow or broad they are; this is referred to as **kurtosis**.

As Figure 2.2 shows, for a normal distribution the mean, median and mode all have the same value. For a data set that is approximately normally distributed, the mean, median and mode will be close to each other. The number of data values distributed below and above this central region tapers off.

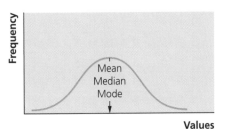

Figure 2.2 A normal distribution curve

The pebble data from Borth Beach summarised in Table 2.6 and Figure 2.1 looks fairly symmetrical. We can therefore say that it approximately follows a normal distribution, which means that the mean, median and mode will all be very similar.

Frequency distribution curves which are not symmetrical are said to be **skewed**.

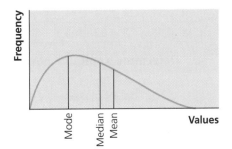

Figure 2.3 A positively skewed distribution curve

Positively skewed distribution

If the frequency distribution curve has a peak on the left side and a longer 'tail' towards the right, like the one in Figure 2.3, it is described as being **positively skewed**. Values will fall mainly below the mean.

Negatively skewed distribution

If the frequency distribution curve has a peak on the right side and a longer 'tail' towards the left, like the one in Figure 2.4, it is described as being **negatively skewed**. The values will fall mainly above the mean. This is the least common type of distribution curve.

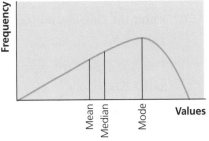

Figure 2.4 A negatively skewed distribution curve

B Guided question

Copy out the workings and complete the answers on a separate piece of paper.

1   Table 2.7 shows data on the percentage cover of marram grass in 160 random quadrat samples on Hawes sand dunes.

Table 2.7 Percentage of marram grass cover for 160 quadrat samples

16	23	20	18	10	20	17	18	22	22
1	13	23	11	10	12	11	20	10	19
7	15	8	13	6	14	8	12	22	9
23	11	16	20	9	20	20	11	8	8
3	5	10	5	20	23	11	18	14	55
19	13	17	5	5	7	17	6	3	2
3	15	9	18	1	37	15	14	21	19
23	20	8	15	15	5	14	22	10	2
3	7	45	2	17	14	16	8	21	17
7	18	1	1	15	10	7	23	16	17
20	21	17	20	10	21	21	7	13	23
12	23	4	19	15	2	14	27	12	12
4	10	12	21	1	32	12	5	10	60
17	15	18	6	9	4	6	22	17	10
50	34	22	4	1	4	17	26	11	9
15	6	22	22	1	15	18	9	10	18

a   Determine a suitable number of classes to group the data into.

$$\text{Number of classes} = \sqrt{\text{total number of data values}}$$

$$= \sqrt{160} = \underline{\quad}, \text{ rounded to } \underline{\quad}$$

b   Determine the width of each class.

Highest data value is ____

Lowest data value is ____

Range of data values is ____ − ____ = ____

Class width = range ÷ number of classes = ____ ÷ ____ = ____, rounded to ____

c   Enter the classes and frequencies into a frequency distribution table as in Table 2.8.

The first class starts from the lowest value and spans an interval of values equal to the class width.

Table 2.8 Frequency table for the marram grass data

Class (% cover)	Frequency
1–5	26

d Draw a histogram to represent the frequency distribution.

The horizontal axis should show the classes.

The vertical axis should cover the range of frequencies from ___ to ___.

e Add a frequency distribution curve to the histogram and interpret it.

The frequency distribution curve is _____, meaning that the data values fall mainly _____ the mean.

C Practice question

2 In a piece of fieldwork to investigate people's perception of place, 20 people were asked to rate each of four aspects of environmental quality (housing, noise, litter, open space) on a scale of 1 to 10. For each person, the four ratings were then added together to give a combined score out of 40. The scores are recorded in Table 2.9.

Table 2.9

2	20	5	23	20	24	34	21	21	18
22	5	20	40	21	21	12	19	23	32

a Determine a suitable number of classes to group the data into.

b Calculate the interval for each class.

c Enter the classes and frequencies into a frequency distribution table.

d Draw a histogram to represent the frequency distribution.

e Add a frequency distribution curve and interpret it.

f Calculate the mean, median and mode and add these values to the graph.

TIP

A quick way to gain a visual impression of the spread of data is to draw a **dispersion diagram**. This consists of an x-axis which is very narrow and will only plot a point in time or a location, and a y-axis on which all of the values in the data set are plotted. The range of the data and any clustering is clearly apparent. Dispersion diagrams are often used as a first step in determining the intervals for a choropleth map.

Geographical application

Fieldwork and the independent investigation

- Measures of central tendency — mean, median and mode — are useful for summarising a data set by a single number.
  - The **mean** value is best for summarising a data set where the values cluster symmetrically around a central value.
  - The **median** is not affected by extreme values and is better than the mean for summarising skewed data sets.
  - The **mode** gives a broad overview of the data set by identifying the most commonly occurring value(s).
- Frequency distributions — in the form of frequency tables, histograms or frequency distribution curves — provide an overall impression of the 'shape' of a data set, such as whether it is symmetrical or skewed.
- If the data can approximately be described by a normal distribution (symmetrical and bell-shaped), then various more complex statistical tests can be applied to the data set for further analysis.

Results interpretation

It is important to remember that frequencies and measures of central tendency, while useful, merely give descriptive summaries of data. As geographers we must use the information they provide to pose further questions in order to gain deeper understanding of the situation being studied.

In the Worked example based on data in **Table 2.1 (page 20)**, on immigration to the UK, by looking at measures of central tendency and the distribution of the data we can make some deductions about migration patterns, such as:

- Between 2005 and 2013 there was little variation in the annual number of immigrants to the UK — the spread of the data for those years is quite small.
- The number of immigrants rose significantly in 2014.

The increase in 2014 raises questions about the origin of the migrants and the various push-and-pull factors that might cause migration, such as geo-political issues, a rise in the minimum wage in the UK and changes in immigration policy.

For further research, we could compare average immigrant numbers to the UK with those in other parts of Europe.

In studies of weather and climate as in **Table 2.2 (page 22)**, measures of central tendency for temperature and rainfall are often used to build up a description of climatic patterns. Departures from these averages can form the basis of discussions about climate change, unseasonal variations and freak weather events.

For the pebble size data in **Table 2.5 (page 25)**, what do the results tell us about the type, size and spatial distribution of pebbles and the sorting of beach material?

For the marram grass data in **Table 2.7 (page 28)**, what factors could explain the distribution of marram grass — soil type, slope, distance from the sea?

For the survey data in **Table 2.9 (page 29)**, what do the results tell us about people's perception of this place? What kind of limitations are there to the data — could any of the four factors which were grouped together in a single score have had a particularly great effect on producing an overall positive or negative impression?

3 Measures of dispersion

Range, interquartile range and standard deviation

Measures of central tendency ('averages' such as the mean, median and mode) are useful for summarising a data set by a single value. However, they tell us nothing about the spread, or **dispersion**, of values. Measures of dispersion are used to represent the degree to which the values in a data set vary from or are dispersed around the average.

For example, two locations may have very similar average annual rainfall amounts, but the rain could occur at different time periods during the year and with very different variations in monthly totals. This might be the case if one location was in an equatorial region while the other had a monsoon climate; the former would experience fairly regular rainfall throughout the year, whereas the latter would have large seasonal variations in rainfall.

Measures of dispersion can give us an idea of whether reliance on the average value might lead to inaccurate or misleading analysis. In geography, the most common measures of dispersion are the **range**, the **interquartile range** and the **standard deviation**.

Range

This is simply the difference between the highest value and the lowest value in a data set. It is easy to calculate, but emphasises the extreme values and does not give any information about the distribution of the remaining values.

Interquartile range and quartile deviation

The interquartile range refers to the interval spanning the middle 50% of values in a data set — that is, 25% of the data on either side of the median. It measures the spread of values around the median.

To calculate the interquartile range, first order the data values and find the median. The median divides the data into two halves. Look at each half of the data (to the left and to the right of the median) and find the median of that half. This will give the **lower quartile** and the **upper quartile**. Then calculate the difference between the upper and lower quartiles; this is the interquartile range. A related and less commonly used measure of dispersion is the **quartile deviation**, which is simply half the interquartile range.

Standard deviation

The standard deviation measures the spread of values around the mean. Unlike the range and interquartile range, it incorporates all the values in the data set. The mathematical formula for standard deviation is

$$\sigma = \sqrt{\frac{\sum (x - \bar{x})^2}{n}}$$

where:

- σ (sigma, the Greek lowercase letter 's') stands for 'standard deviation'.
- Σ (Sigma, the Greek capital letter 'S') stands for 'sum'.
- x is each value in the data set.
- $\bar{x}$ is the mean value of the data set.
- n is the number of items in the data set.

In other words, add up the squared deviations of all the data values from the mean, divide by the total number of data values and then take the square root.

Worked example

Table 3.1 shows the average annual snowfall recorded at 11 weather stations in Alaska.

Calculate the range, interquartile range, quartile deviation and standard deviation of this data.

Table 3.1 Average annual snowfall (centimetres) in Alaska, 1981–2010

Weather station	1	2	3	4	5	6	7	8	9	10	11
Average annual snowfall (cm)	365	665	190	120	160	225	90	170	200	100	300

Source: National Climate Data Center of Alaska

Range

The highest value in the data set is 665 (station 2) and the lowest value is 90 (station 7).

Range = 665 − 90 = 575 cm

Interquartile range

Step 1: arrange the data values in order of size from highest to lowest.
This is done in Table 3.2.

Table 3.2 Data values ordered from high to low

Rank	Average annual snowfall (cm)	
1	665	
2	365	
3	300	Upper quartile
4	225	
5	200	
6	190	Median (middle quartile)
7	170	
8	160	
9	120	Lower quartile
10	100	
11	90	

Step 2: identify the median.
There are 11 values, so the median is the 6th one, which is 190.

Step 3: the median splits the data set into two halves, each consisting of 5 values.
Look at the 5 values above the median. The median of these 5 values, 300, is the upper quartile.
Look at the 5 values below the median. The median of these 5 values, 120, is the lower quartile.

Step 4: calculate the difference between the upper quartile and lower quartile.

Interquartile range = 300 − 120 = 180 cm

Quartile deviation

This is obtained by dividing the interquartile range by 2:

$180 \div 2 = 90\,\text{cm}$

Standard deviation

Step 1: find the mean of the data set, $\bar{x}$.

$365 + 665 + 190 + 120 + 160 + 225 + 90 + 170 + 200 + 100 + 300 = 2585$

$= 2585 \div 11 = 235$

Step 2: subtract the mean from each value in the data set, then square each of these differences to get rid of any minus signs. It helps to construct a table to do these calculations.

Table 3.3 Calculating the standard deviation of the snowfall data

Weather station	Snowfall	Deviation from the mean, $(x - \bar{x})$	Squared deviation from the mean, $(x - \bar{x})^2$
1	365	$365 - 235 = 130$	16 900
2	665	$665 - 235 = 430$	184 900
3	190	$190 - 235 = -45$	2 025
4	120	$120 - 235 = -115$	13 225
5	160	$160 - 235 = -75$	5 625
6	225	$225 - 235 = -10$	100
7	90	$90 - 235 = -145$	21 025
8	170	$170 - 235 = -65$	4 225
9	200	$200 - 235 = -35$	1 225
10	100	$100 - 235 = -135$	18 225
11	300	$300 - 235 = 65$	4 225
			$\sum(x - \bar{x})^2 = 271\,700$

Step 3: add together the values in the last column of Table 3.3.

$\sum(x - \bar{x})^2 = 271\,700$

Step 4: divide this total by n, the number of data values.

$271\,700 \div 11 = 24\,700$

Step 5: take the square root of the value in Step 4 to obtain the standard deviation.

$\sigma = \sqrt{24\,700} = 157\,\text{cm}$

TIP

Once the data has been ordered from high to low, you can also use the following formulae to locate the upper and lower quartiles in the list. (n = number of values in the data set.)

- upper quartile position = $(n + 1) \div 4$
- lower quartile position = $(n + 1) \div 4 \times 3$

So in the Worked example:

- upper quartile position = $(11 + 1) \div 4 = 3$
  upper quartile = 3rd figure from the top of the ranking = 300
- lower quartile position = $(11 + 1) \div 4 \times 3 = 9$
  lower quartile = 9th figure from the top of the ranking = 120
- interquartile range = UQ − LQ. 300 cm − 120 cm = 180 cm

The smaller the interquartile range or quartile deviation, the narrower the spread or dispersion from the median, i.e. the stronger the clustering around the middle value. The larger the interquartile range or quartile deviation, the greater the spread or dispersion from the median, i.e. the weaker the clustering around the middle value.

Similarly, the size of the standard deviation indicates the degree of spread or dispersion from the mean value.

For the snowfall data in the Worked example, all the measures of dispersion are quite high, indicating that the data is widely dispersed from both the median and the mean. This is therefore an example of a data set for which an 'average' value cannot be relied on to accurately reflect the data.

A geographer would undertake further investigations to seek explanations for the wide spread of values. For example, the locations of the weather stations are an obvious starting point, as stations in northern Alaska would presumably have larger snowfalls.

B Guided question

Copy out the workings and complete the answers on a separate piece of paper.

1 **The data in Table 3.4 shows moisture readings for 11 soil samples taken at a deciduous woodland location. (The scale is from 1 to 10, where 1 is dry and 10 is moist.)**

Table 3.4 Soil samples from a woodland site

Sample	Soil moisture reading
1	6.5
2	7
3	7
4	5.5
5	8
6	4.8
7	6.8
8	8
9	6
10	9
11	7

a **Calculate the range.**

The highest value in the data set is _____ and the lowest value is _____

Range = _____ − _____ = _____

b Calculate the interquartile range and quartile deviation.

Step 1: order the data values from high to low. (Copy and fill in Table 3.5.)

Step 2: find the upper quartile.

Using the formula:

Upper quartile position = $(11 + 1) \div 4 = 3$
Upper quartile = 3rd figure from the top of the ranking = _____

Step 3: find the lower quartile.

Lower quartile position = $(11 + 1) \div 4 \times 3 = 9$
Lower quartile = 9th figure from the top of the ranking = _____

Step 4: calculate the difference between the upper quartile and lower quartile.

Interquartile range = _____ − _____ = _____

Step 5: the quartile deviation is half the interquartile range.

Quartile deviation = _____ $\div 2$ = _____

This is a large/small number, indicating a high/low degree of dispersion around the median value.

Table 3.5 Soil moisture readings ordered from high to low

Soil moisture reading
9
4.8

c Calculate the standard deviation.

Step 1: find the mean of the data set, $\bar{x}$.

$(6.5 + \underline{\ \ } + \underline{\ \ } + \underline{\ \ } + \underline{\ \ } + \underline{\ \ } + \underline{\ \ } + \underline{\ \ } + \underline{\ \ } + \underline{\ \ } + \underline{\ \ }) \div 11 = \underline{\ \ }$

Step 2: subtract the mean from each value in the data set and then square each of these differences. (Fill in a copy of Table 3.6 for these calculations.)

Table 3.6

Sample	Soil moisture	Deviation from the mean, $(x-\bar{x})$	Squared deviation from the mean, $(x-\bar{x})^2$
1	6.5	−0.5	0.25
2	7		
3	7		
4	5.5		
5	8		
6	4.8		
7	6.8		
8	8		
9	6		
10	9		
11	7		
		Sum of squared deviations:	

Step 3: add together the values in the last column of Table 3.6.

$\Sigma(x - \bar{x})^2 =$ _____

Step 4: divide this total by n, the number of data values.

_____ ÷ 11 = _____

Step 5: take the square root of the value obtained in Step 4 to obtain the standard deviation.

$$\sigma = \sqrt{} = \underline{}$$

This is a high/low value, indicating that the data is/is not widely dispersed from the mean.

ⓒ Practice question

2

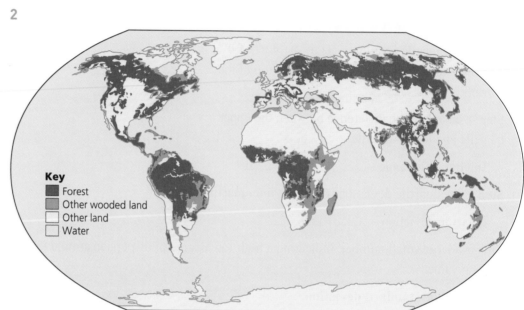

Figure 3.1 Map of the world's forests

Source: Food and Agriculture Organization of the United Nations (FAO)

Table 3.7 shows forest areas as a percentage of total land area for a selection of advanced countries (ACs) and low-income developing countries (LIDCs).

Table 3.7 Forest area of selected ACs and LIDCs in 2015

AC	Forest area (% total land area)	LIDC	Forest area (% total land area)
Austria	47	Bangladesh	11
Canada	38	Burundi	10
France	32	Cameroon	41
Finland	73	Ethiopia	12
Germany	33	Ghana	42
Italy	31	Malawi	34
Japan	69	Nepal	25
New Zealand	39	Sudan	8
Portugal	35	Tanzania	53
Spain	37	Vietnam	47
UK	13	Zambia	66

Source: World Bank

a   Compare the ranges of the AC and LIDC data sets.
b   For the two data sets, calculate and compare the
    i   interquartile ranges
    ii   quartile deviations
c   Calculate the standard deviation for the LIDCs.
d   Interpret the standard deviation value from part **c**.

TIP

In this chapter, all the example data sets have contained an odd number of values. When you have a data set with an even number of values, here's how to find the interquartile range. For example, consider the following set of 10 values (already ordered from high to low):

21   20   16   15   11 | 9   8   7   5   3

Because there is an even number of values, you can split the data set into two equal halves down the middle, indicated by the red vertical line.

The median (middle value) of the values in the left half will be the upper quartile — 16 in this case.

The median (middle value) of the values in the right half will be the lower quartile — 7 in this case.

Finally, calculate the difference between the upper and lower quartiles to obtain the interquartile range: $16 - 7 = 9$.

If each half of the data set also contains an even number of values, the upper or lower quartile will be the midpoint between the two centre values of that half.

Geographical application

Fieldwork and the independent investigation

- Two sets of data can have the same average (measure of central tendency) but very different spreads of values. Measures of dispersion help to quantify the scatter of values in a data set and indicate how reliable the average is as a representative value of the whole data set.
- The interquartile range or quartile deviation measures how widely dispersed the data values are from the median.
- The standard deviation measures how widely dispersed the data values are from the mean. It is one of the most important descriptive statistics for research and fieldwork because:
  - It is calculated using all the values in the data set.
  - It is a fundamental ingredient of more complex 'inferential' statistical tests for investigating the significance of relationships and differences between data sets.

  A low standard deviation indicates that most of the values are close to the mean, whereas a high standard deviation tells us that the mean is not so reliable for summarising the whole data set.
- In fieldwork, measures of dispersion can also be used to support (or contradict) an argument or hypothesis.

Results Interpretation

For the data in **Table 3.1 (page 32)**, geographers might be interested in examining the locational factors which could affect the amount of snowfall recorded at different weather stations in Alaska over various time periods.

For the data in **Table 3.4 (page 34)**, geographers might want to investigate factors affecting soil moisture, such as slope, soil texture and climate.

The data in **Table 3.7 (page 36)** reveals distinct differences in forest cover between different LIDCs and also between LIDCs and ACs. Agricultural practices, food security issues, legislation on forest removal, climate and habitat protection are all possible factors contributing to these differences.

4 Measures of concentration

Nearest neighbour index

In geography, many practical investigations involve the interpretation of spatial pattern. The data is usually displayed on a map initially, but in order to interpret and compare patterns numerically, a statistical tool called **nearest neighbour analysis** can be used.

Originally used by ecologists to describe patterns of plant distribution, nearest neighbour analysis produces a value, the **nearest neighbour index (NNI)**, which measures the extent to which a spatial distribution is 'clustered'.

The NNI is usually denoted by 'Rn' and the formula used to calculate it is

$$\text{Rn} = 2\bar{d}\sqrt{\frac{n}{A}}$$

where:

- $\bar{d}$ is the mean nearest neighbour distance.
- n is the total number of points in the survey.
- A is the study area.

Rn values can range from 0 to 2.15 and have the following interpretations (see Figure 4.1).

- Rn = 0: **clustered** — a pattern where the points are grouped closely together. For example, specialist shopping areas like the jewellery quarter in Birmingham will exhibit this kind of pattern.
- Rn = 1.0: **random** — points are randomly distributed throughout the area.
- Rn = 2.15: **regular** — a pattern which is perfectly uniform with regular spacing, such that all the points are at an equal distance from each other. In reality this rarely occurs; the nearest example might be the distribution of settlements in a relatively undifferentiated landscape, such as flat agricultural land like that in East Anglia or the midwestern USA.

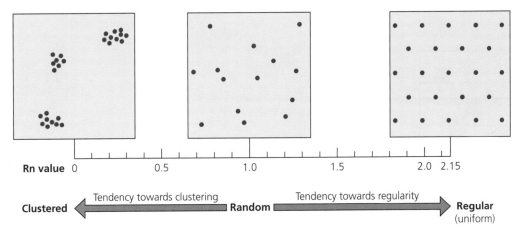

Figure 4.1 Nearest neighbour index values and their interpretations

Worked example

Figure 4.2 shows the locations of cafes and outdoor clothing stores in a small town in the Lake District. Calculate and interpret the NNI for the cafes.

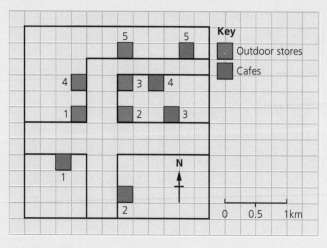

Figure 4.2

Step 1: calculate the study area, A.

The area mapped consists of 12×12 squares on the grid and the scale shows that the length of 4 squares represents 1 km, so the study area is

$$A = 3\,\text{km} \times 3\,\text{km} = 9\,\text{km}^2$$

Step 2: plot the locations of the cafes and number them.

In Figure 4.2 the cafes are shown in green and have already been numbered 1 to 5.

Step 3: measure the distance from each cafe to its nearest neighbour and make a table to summarise this information.

- It does not matter if one location is the nearest neighbour of several other locations.
- There are many ways of defining 'distance'. For simplicity we will take the distance between two locations to be the maximum horizontal or vertical gap between them (i.e. do not worry about measuring diagonally across the squares in the grid).

The distances are plotted in Figure 4.3 and entered in Table 4.1.

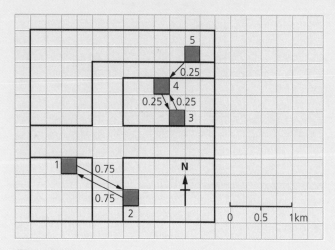

Figure 4.3

Full worked solutions at **www.hoddereducation.co.uk/essentialmathsanswers**

Table 4.1 Nearest neighbour distances between the cafes

Cafe number	Nearest neighbour number	Distance, d (km)
1	2	0.75
2	1	0.75
3	4	0.25
4	3	0.25
5	4	0.25
	Sum of the distances:	$\sum d = 2.25$

Step 4: calculate the mean nearest neighbour distance, $\bar{d}$.

This is obtained by adding together all the nearest neighbour distances d and dividing by the number of locations (n):

$$\bar{d} = \sum d \div n = 2.25 \div 5 = 0.45$$

Step 5: insert the values for $\bar{d}$, n and A into the NNI formula and calculate the Rn value.

$$Rn = 2\bar{d}\sqrt{\frac{n}{A}}$$

$$= 2 \times 0.45 \times \sqrt{\frac{5}{9}} = 0.9 \times 0.745 = 0.67$$

Step 6: interpret the NNI.

The Rn value of 0.67 is below 1.0 and indicates that the cafes have a tendency towards clustering (see Figure 4.1).

B Guided question

Copy out the workings and complete the answers on a separate piece of paper.

1 a **Calculate the NNI for the outdoor shops in Figure 4.2.**

Step 1: calculate the study area, A.

This is the same as in the Worked example: $A =$ _____ km²

Step 2: plot the locations of the outdoor shops and number them.

In Figure 4.2 the outdoor shops are shown in red and have already been numbered 1 to 5.

Step 3: measure the distance from each outdoor shop to its nearest neighbour and summarise this information in a copy of Table 4.2. (Use the same way of measuring distances as in the Worked example.)

Table 4.2 Nearest neighbour distances between the outdoor shops

Outdoor shop number	Nearest neighbour number	Distance, d (km)
1	4	0.25
2		
3		
4		
5		
	Sum of the distances:	$\sum d =$

Step 4: calculate the mean nearest neighbour distance, $\bar{d}$.

Add together the nearest neighbour distances d and divide by the number of locations (n):

$$\bar{d} = \sum d \div n = \underline{\hspace{1cm}} \div 5 = \underline{\hspace{1cm}}$$

Step 5: insert the values for $\bar{d}$, n and A into the NNI formula and calculate the Rn value.

$$Rn = 2\bar{d}\sqrt{\frac{n}{A}}$$

$$= 2 \times \underline{\hspace{1cm}} \times \sqrt{\frac{5}{\Box}}$$

$$= \underline{\hspace{1cm}}$$

Step 6: interpret the NNI.

The Rn value of _____ is below/above 1.0 and indicates that the outdoor shops have a weak/strong tendency towards _____.

b Compare the distributions of the cafes and the outdoor shops. Which is the more clustered?

The _____ have a smaller Rn value, which indicates that they are more clustered than the _____.

c Suggest possible reasons for your result in part b.

This might be because _____

Ⓒ Practice question

2 As part of a mapwork activity, students plotted the locations of water-filled corries at two contrasting locations in the UK, Snowdonia and Langdale Fell; see Figures 4.4 and 4.5.

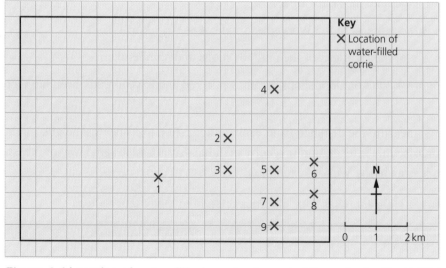

Figure 4.4 Location of water-filled corries in part of Snowdonia, North Wales

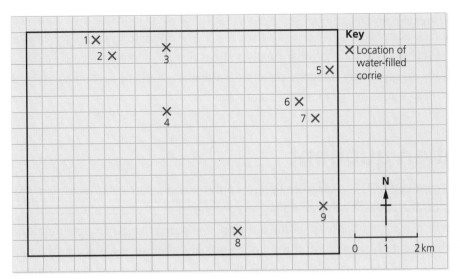

Figure 4.5 Location of water-filled corries in Langdale Fell, Cumbria

a   Calculate and interpret the NNI for the water-filled corries in Snowdonia, North Wales.

b   Calculate and interpret the NNI for the water-filled corries in Langdale Fell, Cumbria.

c   Which is the more clustered?

d   What further information would be needed for a possible explanation of the results?

TIP

Nearest neighbour analysis is most informative and meaningful when it is used to **compare** distributions, rather than being applied to one data set only. When comparing spatial distributions, it is important that each data set covers an area of the same size; in other words, the value of A in the formula should be the same for the data sets being compared. The size of the study area can affect the Rn value considerably.

Critical values tables can be used for Nearest neighbour in order to confirm if significant clustering or dispersal has occurred.

n is the number of points in the survey. In the Worked example on page 40, the NNI was 0.67 and $n = 5$.

n	Clustered pattern		Dispersed pattern	
	0.05	0.01	0.05	0.01
5	0.616	0.456	1.385	1.544

A value less than 0.616 is considered to indicate significant clustering and a value greater than 1.385 is considered to indicate significant dispersal. Because of the proximity of 0.67 to 0.616, we can again conclude a tendency towards clustering.

Be aware that when the NNI indicates a clustered distribution (i.e. the Rn value is towards the zero end of the scale), it cannot distinguish between a single cluster and a multi-clustered pattern. Figure 4.6 shows an example of a single-clustered configuration and a multi-clustered distribution that may have the same or similar Rn values.

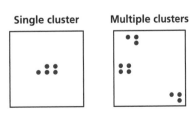

Figure 4.6 A single cluster and a multi-clustered distribution

In geography, when the NNI indicates a random distribution, it is important to think about underlying factors that could give rise to such a result. For example, a seemingly 'random' pattern of villages may be due to water supply sources being randomly spaced over the landscape; and an apparently 'random' pattern of vegetation could be accounted for by the distribution of a controlling factor such as soil type.

Location quotient

A location quotient (LQ) is a measure of how concentrated a particular factor is within a certain area. In geography this could be the concentration of a particular industry, occupation or demographic group within a region, compared to a larger geographical area, such as nationally. It is a useful way of identifying what gives a place its unique characteristics and the information provided by LQs can be used to compile a **place profile**.

A location quotient is an index number, defined using fractions or ratios. For instance, if employment is the factor being studied, the LQ is given by the formula

$$LQ_i = \frac{e_i/e}{E_i/E}$$

where:

- LQ_i denotes the location quotient for industry i in the regional economy.
- e_i is employment in industry i in the regional economy.
- e is total employment in the region.
- E_i is employment in industry i in the national economy.
- E is total employment nationally.

When the ratios $\frac{e_i}{e}$ and $\frac{E_i}{E}$ are expressed as percentages, the formula can be simplified to

$$LQ_i = \frac{\% e_i}{\% E_i}$$

In other words, the percentage regional employment in industry i is divided by the percentage national employment in industry i.

Location quotients are interpreted as follows:

- A value of **1** means that in the location of interest, the factor is represented with exactly the same concentration as in the larger geographical area.
- A value **greater than 1** means that in the location of interest, the factor has a higher concentration than average (i.e. over the larger geographical area).
- A value **less than 1** means that in the location of interest, the factor has a lower than average concentration.

The geographical areas of comparison can vary — for example, regional compared to national scale, or local (an individual town or city) compared to regional scale.

Although the main use of LQs in geography is to compare economic variables such as employment rates in different parts of the economy (see the Worked example below), many other factors can be studied, including population demographics (proportions of retired or working-age people, percentages of the population with various levels of qualifications etc.) and a whole range of characteristics from census data (e.g. proportions of the population belonging to different cultural, religious, ethnic and social groups).

A Worked example

Table 4.3 shows employment rates in three industrial sectors in the North West region compared to the UK nationally. For each sector, calculate and interpret the location quotient of employment.

Table 4.3 Employment in selected industrial sectors in the North West region and the UK nationally, 2015

Industrial sector	% employment in NW region	% employment in the UK nationally
Manufacturing	9.7	7.8
Information & communications	2.8	4.0
Health	14.4	12.4

Source: Office for National Statistics (ONS)

Because the employment rates are given as percentages, we can use the simpler formula

$$LQ_i = \frac{\% \, e_i}{\% \, E_i}$$

Applying this formula to each of the sectors:
- In manufacturing, LQ = 9.7 ÷ 7.8 = 1.2 (> 1).
- In information & communications, LQ = 2.8 ÷ 4.0 = 0.7 (< 1).
- In health, LQ = 14.4 ÷ 12.4 = 1.2 (> 1).

Therefore, the NW region shows a higher-than-national-average concentration of employment in the manufacturing and health sectors, but a lower-than-national-average concentration of employment in information & communications.

A natural extension of the investigation in the Worked example is to compare different regions, which is done in the Guided question below.

B Guided question

Copy out the workings and complete the answers on a separate piece of paper.

1 Complete a copy of Table 4.4 by filling in the values in the LQ columns.

Table 4.4 Employment in three industrial sectors in seven regions of the UK, 2015

	Manufacturing		Information & communications		Health	
	% employment	LQ	% employment	LQ	% employment	LQ
UK	**7.8**	N/A	**4.0**	N/A	**12.4**	N/A
North West (NW)	9.7	1.2	2.8	0.7	14.4	1.2
North East (NE)	10.5		2.4		16.7	
South West (SW)	8.1		3.2		13.1	
South East (SE)	5.7		5.4		10.8	
Scotland	7.4		2.5		14.3	
Wales	10.6		2.3		14.2	
Northern Ireland	10.0		2.4		15.7	

Source: Office for National Statistics (ONS)

The values for the NW region, calculated in the Worked example, have been filled in already.

The other entries are calculated similarly. For example:

LQ for manufacturing in the NE region = 10.5 ÷ 7.8 = _____

LQ for health in the SW region = 13.1 ÷ _____ = _____

LQ for information & communications in Scotland = _____ ÷ 4.0 = _____ and so on.

Ⓒ Practice questions

Calculate the LQ value for the comparison on each of the spatial scales given in Table 4.5.

Table 4.5 Comparison of population demographics on different spatial scales, 2015

2 Scale: individual settlement compared to national		
	Retired people	
	%	LQ
Great Britain	14.1	N/A
Bamburgh (Northumberland)	24.6	

3 Scale: county compared to regional		
	Retired people	
	%	LQ
South West region	19.0	N/A
Devon	25.4	

4 Scale: individual settlement compared to national		
	Students	
	%	LQ
Great Britain	26.2	N/A
Cambridge	46.3	

5 Scale: county compared to national		
	Managing directors	
	%	LQ
Great Britain	10.3	N/A
Surrey	13.0	

Source: Office for National Statistics (ONS)

Lorenz curves

A Lorenz curve is a graph used to show the concentration of a particular factor or variable — such as population, industry, employment or land use — over different areas. These 'areas' could be geographical locations, or other categories such as age groups, industrial sectors or income levels.

To plot a Lorenz curve, fieldwork data can be used, though it is more common to use secondary data. The procedure is as follows.

Step 1: collect data on the variable and areas you are interested in — that is, find out the values of the variable for the different areas.

Step 2: convert the values to **percentages**. For example, if 22 people work in sales, out of a total workforce of 150, then the percentage is 22 ÷ 150 × 100% = 14.7%.

Step 3: **order** the percentages. Keep each one together with the area it corresponds to; then the areas will also be ranked in order of their relative importance.

Step 4: go down the rank order and calculate the **cumulative** (running total of) percentages.

Step 5: make a graph by plotting the rank order (or the ranked variable) on the horizontal (*x*) axis and the cumulative percentages on the vertical (*y*) axis.

Step 6: add an **even distribution line** to represent the pattern of distribution if all the areas had an equal percentage.

The graph produced in Step 5 is a Lorenz curve. The further the Lorenz curve is from the even distribution line of Step 6, and the more concave its shape, the more unevenly the data is distributed among the different areas — in other words, the greater the **concentration** of the factor in certain areas. Lorenz curves can lie above or below the even distribution line.

Several Lorenz curves can be plotted on the same graph, as in Figure 4.7, which shows Lorenz curves for employment in three different sectors over ten geographical regions. This allows comparisons to be made between areas or factors.

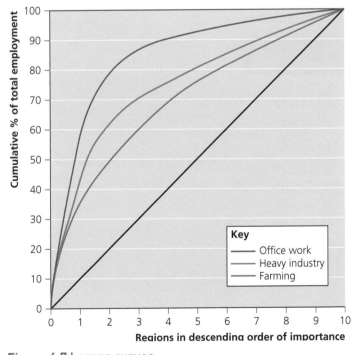

Figure 4.7 Lorenz curves

Worked example

The secondary data in Table 4.6 was collected to compare the employment types in three locations. (Here 'elementary labour' refers to work in a primary industry, such as farming and fishing, or of a basic technical level, such as cleaning and delivery.)

Table 4.6 Employment types as percentages of total employment in three UK locations, 2014

Employment type	Consett, County Durham % of total employment	Exeter, Devon % of total employment	Alnwick, Northumberland % of total employment
Administration	10	6	7
Elementary labour	14	33	18
Machinery	16	22	6
Managers	7	4	11
Personal services	9	9	12
Professionals	13	3	14
Sales	8	8	9
Technical	12	5	10
Trades	11	10	13

Source: Office for National Statistics (ONS)

Plot Lorenz curves for the three locations on the same graph and interpret the results.

Steps 1 and 2 of the procedure described above have been completed, as the data is already collected and presented in the form of percentages.

Step 3: for each location, order the percentages from high to low. Keep each percentage together with the area of employment it corresponds to; then the employment types will also be ranked in order of their relative importance.

It is best to do this by making a table for each location. See Tables 4.7–4.9.

Table 4.7 Ranked employment data for Consett, County Durham

Employment type	% of total employment	Rank	Cumulative %
Machinery	16	1	16
Elementary labour	14	2	16 + 14 = 30
Professionals	13	3	30 + 13 = 43
Technical	12	4	43 + 12 = 55
Trades	11	5	55 + 11 = 66
Administration	10	6	66 + 10 = 76
Personal services	9	7	76 + 9 = 85
Sales	8	8	85 + 8 = 93
Managers	7	9	93 + 7 = 100

Table 4.8 Ranked employment data for Exeter, Devon

Employment type	% of total employment	Rank	Cumulative %
Elementary labour	33	1	33
Machinery	22	2	55
Trades	10	3	65
Personal services	9	4	74
Sales	8	5	82
Administration	6	6	88
Technical	5	7	93
Managers	4	8	97
Professionals	3	9	100

Table 4.9 Ranked employment data for Alnwick, Northumberland

Employment type	% of total employment	Rank	Cumulative %
Elementary labour	18	1	18
Professionals	14	2	32
Trades	13	3	45
Personal services	12	4	57
Managers	11	5	68
Technical	10	6	78
Sales	9	7	87
Administration	7	8	94
Machinery	6	9	100

Step 4: for each location, go down the rank order and calculate the cumulative (running total of) percentages. These cumulative percentages have been entered into an extra column in each of Tables 4.7–4.9.

Step 5: for each location, plot the rank order of employment type on the horizontal (x) axis and the cumulative percentage of total employment on the vertical (y) axis.

The curves for the three locations can all be displayed on the same graph, as shown in Figure 4.8.

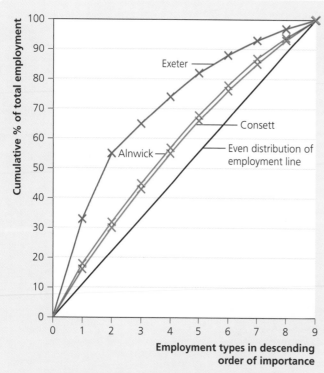

Figure 4.8 Lorenz curves for employment types in three UK towns

Step 6: add an even distribution line to represent the pattern of distribution if all the employment types had an equal percentage. This is the diagonal straight line in Figure 4.8.

Interpretation of the graph may include the following points:

- By ranking the data for each location, the same numbers (the rank orders) can be plotted on the x-axis for all the curves.

- By using percentages rather than raw employment figures, differences in the working population sizes of the three locations can be accounted for, allowing the 'standardised' data sets to be compared with each other.

- If the same proportion of people were employed in each category, the curves would coincide with the even distribution line.

- All three Lorenz curves show deviation from 'even distribution of employment'. The greater the distance from the even distribution line and the more concave the curve, the greater the concentration of employment in certain categories.

- Exeter displays the greatest degree of concentration, but it is necessary to refer back to Table 4.8 to see exactly which categories account for the largest proportion of employment (elementary labour and machinery).

- Alnwick and Consett have similar degrees of concentration, but Tables 4.7 and 4.9 show that the concentration is in different categories of employment for the two locations.

B Guided question

Copy out the workings and complete the answers on a separate piece of paper.

1 **A frequent use of Lorenz curves in geography is to study the distribution of population. Figure 4.9 shows the regions of England and Table 4.10 contains data on the population and land area of the regions.**

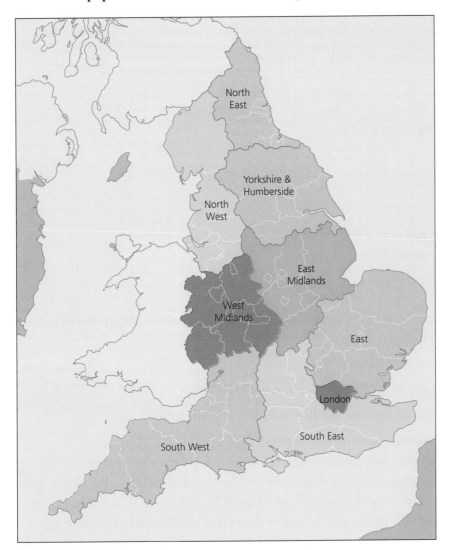

Figure 4.9 Regions of England

Table 4.10 Population of regions in England, 2014

Region	Population (millions)	Area (km²)
South East (SE)	8.6	19 095
London	8.1	1 572
North West (NW)	7.1	14 165
East (E)	5.9	19 192
West Midlands (WM)	5.7	13 000
South West (SW)	5.3	23 829
Yorkshire & Humberside (Y&H)	5.2	15 420
East Midlands (EM)	4.5	15 627
North East (NE)	2.6	8 592
Total	**53.0**	**130 492**

Source: Office for National Statistics (ONS)

a Complete Steps 2–4 of constructing a Lorenz curve (see page 47) for this data and fill in a copy of Table 4.11.

Table 4.11 Percentage populations of regions in England, ordered from highest to lowest

Step 3: Region and rank (in descending order of % population)	Step 2: % population of England	Step 4: Cumulative % of population	Step 2: % land area of England	Step 4: Cumulative % of land area
South East (SE)	$8.6 \div 53 \times 100$ = 16	16	$19\,095 \div 130\,492 \times 100 = 15$	15
London	$8.1 \div 53 \times 100$ = 15	16 + 15 = 31		
North West (NW)				
East (E)				
West Midlands (WM)				
South West (SW)				
Yorkshire & Humberside (Y&H)				
East Midlands (EM)				
North East (NE)				

b Complete Step 5 (see page 47) — construct a Lorenz curve from the data in your completed Table 4.11.

Plot the ranked variable (cumulative percentage of population) on the *x*-axis and cumulative percentage of land area on the *y*-axis.

c Add an even distribution line and interpret the graph.

d Interpret the graph using the following questions for guidance:
   ▪ Does the Lorenz curve indicate a high or low degree of concentration?
   ▪ Which regions account for most of the concentration? (In other words, which region corresponds to the furthest point from the even distribution line?)
   ▪ Using supporting evidence from your completed Table 4.11, do the largest regions by land area have the highest proportion of population?

C Practice question

2 a Following Steps 3 and 4 for constructing Lorenz curves, complete a copy of Table 4.12.

Table 4.12 Wealth distribution in the UK, Canada and Belgium

Cumulative % of population	UK		Belgium		Canada	
	% wealth held by each 10% of population from poor to rich	Cumulative % of wealth	% wealth held by each 10% of population from poor to rich	Cumulative % of wealth	% wealth held by each 10% of population from poor to rich	Cumulative % of wealth
(Poorest 10%) 10	1		3		0.5	
20	2		4		1	
30	3		5		1.5	
40	4		7		2	
50	5		8		4	
60	6		10		6	
70	8		12		7	
80	12		15		10	
90	18		16		20	
100 (Richest 10%)	41		20		48	

Source: OECD Income Distribution database

b Following Steps 5 and 6, draw three Lorenz curves on the same graph to compare the distribution of wealth in the UK, Belgium and Canada.

c Add an even distribution line to the graph. Which country shows the lowest degree of wealth concentration across its population?

Geographical application

Fieldwork and the independent investigation

- The **nearest neighbour index** was developed by a botanist to describe patterns of plant distribution. It can be used in fieldwork to describe various spatial distributions, most commonly of settlements or shops, but also of types of industry or physical geographical features. It must be remembered that the nearest neighbour index only indicates the degree of clustering, randomness or regularity of the distribution, but provides **no explanation** of the pattern, so further work must be carried out on the physical and human geography of the study area to understand the factors that might have given rise to the pattern.

- **Location quotients** measure how concentrated a particular factor is within a certain area. The results can be used to compare and contrast different areas. In terms of data presentation, location quotient values can be mapped, for instance on a **choropleth map** (see Figure 4.10). The pattern on the map can then be described and explained, drawing on geographical knowledge. For Figure 4.10 this may be factors such as infrastructure, resources, skills, population structure and industrial inertia.

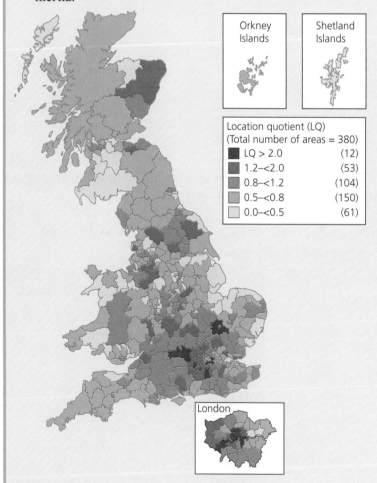

Figure 4.10 Choropleth map showing manufacturing employment in Great Britain, 2013

- **Lorenz curves** provide a graphical representation of the concentration of a factor — such as population, wealth, economic activity in particular sectors and employment types — over certain areas or categories. Several curves can be plotted on one graph, allowing comparisons to be made between spatial locations on a range of scales (global, national, regional and local) or over time. Note, however, that the sizes of the areas or categories should be broadly similar, or else deviations from 'even distribution' will be exaggerated.

All the examples and questions in this chapter have made use of secondary data, such as census figures, maps and government statistics, but the techniques are also applicable to primary data collected from fieldwork.

Results interpretation

For the data in **Table 4.5 (page 46)**, what factors could account for high concentrations of retired people in Northumberland and Devon, e.g. coastal locations, and for managing directors in Surrey, e.g. proximity to London?

For the Lorenz curve based on the data in **Table 4.12 (page 53)**, what factors may lead to a high concentration of wealth in a small fraction of the population? Consider the job opportunities, education and skill levels and political factors in the different countries. Why might it be that a country such as Belgium has much less income inequality than the UK or Canada?

5 Measures of correlation

Correlation refers to the degree of **association** between two sets of data or two variables. In geography fieldwork, we often seek to examine the relationship between factors — how does one factor affect another (if at all)? For example, a geographer might investigate the influence of velocity on the rates of abrasion in glaciers, the effect of distance from the central business district on the cost of land in urban areas, or the relationship between energy use and gross national income in selected countries.

A basic analysis of the correlation between two sets of data (the values of two variables or factors) proceeds in the following steps.

Step 1: draw a **scatter graph**.

Step 2: add a **line of best fit** to the scatter graph.

Step 3: calculate the **correlation coefficient** for the pair of data sets.

Step 4: compare the value of the correlation coefficient with a **critical value** to assess the **significance** of the relationship between the two factors and make a **concluding comment**.

Scatter graphs and lines of best fit

Initially, a **scatter graph** is used to give an idea of the **likely relationship** between the two factors. If it is thought that one factor (the **independent variable**) may have an effect on the other factor (the **dependent variable**), plot the values of the independent variable on the horizontal (x) axis and the corresponding values of the dependent variable on the vertical (y) axis.

A **line of best fit** can then be added to highlight the general trend of the data. The line can be calculated mathematically, but in geography it is usually enough to draw one 'by eye'. As far as possible, make sure there are an equal number of points above and below the line. The line of best fit does not have to start at (0, 0) on the graph. Figure 5.1 shows several scatter graphs and describes the types of correlation they represent.

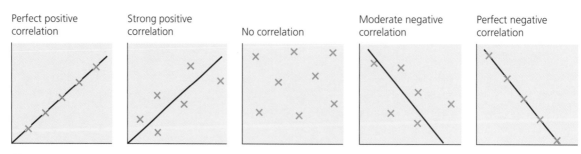

Figure 5.1 Types of correlation on a scatter graph with a line of best fit

Worked example

Table 5.1 shows the gross national income (GNI) per capita and the health expenditure per capita of selected countries in 2014 (all figures are in US dollars, rounded up to nearest thousand).

Table 5.1 GNI per capita and health expenditure per capita of selected countries in 2014

Country	GNI per capita	Health expenditure per capita
Australia	65 000	6 000
Norway	104 000	10 000
Japan	40 000	4 000
Switzerland	88 000	9 000
Spain	29 000	3 000
Chile	15 000	1 000
Mexico	10 000	700
Russia	13 000	1 000
Peru	6 000	400
China	7 000	400
Botswana	7 000	400
Bangladesh	1 000	30
Ghana	2 000	100
Uganda	1 000	60
Nepal	700	40

Source: World Bank

Draw a scatter graph and line of best fit for the data.

We expect health expenditure to be the dependent variable, as it should be affected by a country's GNI. So plot the independent variable, GNI, on the *x*-axis and health expenditure on the *y*-axis.

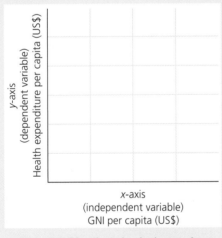

Figure 5.2 Plotting the independent and dependent variables on a scatter graph

Plot each pair of data values on the graph as a scatter point, marked with a cross. In Figure 5.3 the country names have been added for ease of analysis.

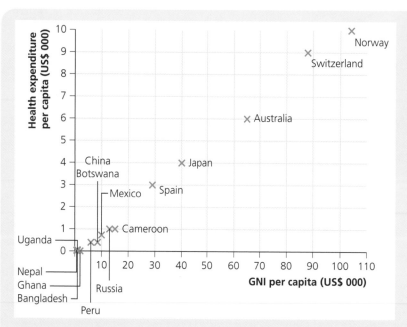

Figure 5.3 Scatter graph of GNI and health expenditure in selected countries

To identify the trend in the data, add a line of best fit to the scatter graph, as shown in Figure 5.4.

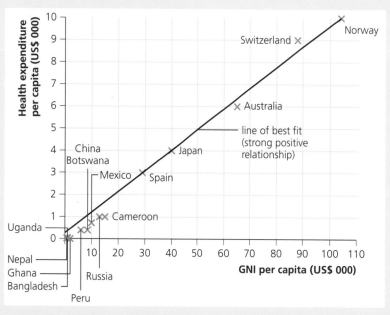

Figure 5.4 Scatter graph with line of best fit to show the relationship between GNI and health expenditure

From Figure 5.4 it can be seen that there is a positive correlation between a country's GNI and its health expenditure: as GNI increases, health expenditure also tends to increase.

The possible reasons for this relationship should then be explored in further analysis.

As well as giving us an initial impression of the relationship between two factors, scatter graphs also allow us to easily identify **anomalies** (sometimes referred to as 'exceptions' or 'outliers'). These are points that lie far off the line of best fit. You can circle them in the scatter graph and in your later geographical analysis of the results it is important to attempt to account for these exceptional data points.

There are no obvious anomalies in the above example.

B Guided question

Copy out the workings and complete the answers on a separate piece of paper.

1   **Consider the data in Table 5.2.**

Table 5.2 GNI per capita (in US dollars rounded up to nearest thousand) and percentage of territorial waters protected in selected countries

Country	GNI per capita	% of territorial waters protected
Australia	65 000	33
Norway	104 000	3
Japan	40 000	6
Chile	15 000	5
Mexico	10 000	19
Russia	13 000	12
Peru	6 000	4
Bangladesh	1 000	3
Ghana	2 000	2
Kenya	1 000	11

Source: World Bank

a   **Draw a scatter graph of the data.**

The axes have been drawn on Figure 5.5 showing the dependent and independent variables.

The independent variable to plot on the *x*-axis is _____.

The dependent variable to plot on the *y*-axis is _____.

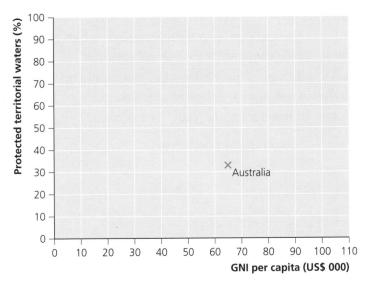

Figure 5.5 Scatter graph of GNI and protected territorial waters

b   **Add a line of best fit to the scatter graph.**

c   **Write a statement summarising the relationship between GNI per capita and percentage of protected territorial waters in the ten selected countries.**

From the scatter graph it appears that there is _____ correlation between a country's GNI per capita and the percentage of its territorial waters that are protected.

ⓒ Practice question

2 Consider the data in Table 5.3.

 a Draw a scatter graph.

 b Add a line of best fit to the scatter graph.

 c Write a statement summarising the relationship shown by the scatter graph.

Table 5.3 GNI per capita (in US dollars rounded up to nearest thousand) and percentage employment in agriculture of selected countries in 2014

Country	GNI per capita	% employment in agriculture
Australia	65 000	3
Norway	104 000	2
Japan	40 000	4
Chile	15 000	10
Mexico	10 000	13
Russia	13 000	8
Peru	6 000	26
Bangladesh	1 000	63
Ghana	2 000	45
Uganda	1 000	72

Source: World Bank

Spearman rank correlation coefficient

In the previous Worked example, the scatter graph of GNI per capita and health expenditure per capita suggests a positive correlation between the two factors, but how strong is the relationship? And is it 'statistically significant' or could it have occurred just by chance? The **Spearman rank correlation coefficient** provides a numerical value that indicates the nature (positive, negative or none) and strength of the relationship. Then a statistical significance test, the **Spearman rank correlation test**, can tell us how reliable the result is — that is, whether the relationship is significantly different from what could have occurred by chance.

The formula for Spearman's rank correlation coefficient is

$$R_S = 1 - \frac{6\sum d^2}{n^3 - n}$$

where:

- d is the difference in rank order of each pair of data values.
- n is the number of pairs of data values.

The R_S value is usually calculated to two decimal places and is interpreted as follows.

- A positive value indicates positive correlation: as one factor increases, so does the other. A value of +1 indicates perfect positive correlation — the data points lie exactly on a line with positive slope.
- A negative value indicates negative correlation: as one factor increases, the other decreases. A value of −1 indicates perfect negative correlation — the data points lie exactly on a line with negative slope.
- A value of 0 (or close to 0) indicates no correlation.

The closer R_S is to +1 or −1, the stronger the likely correlation between the two factors.

Worked example

Calculate the Spearman rank correlation coefficient for the data in Table 5.1 on GNI per capita and health expenditure per capita of 15 selected countries in 2014 and use the Spearman rank test to determine whether the result is statistically significant.

Step 1: a statistical test always begins with a **null hypothesis**.
For the Spearman rank test this is a written assumption that there is no relationship:

There is no relationship between a country's GNI per capita and health expenditure per capita.

Step 2: rank both sets of data values.
It does not matter whether you rank from high to low or from low to high, but whichever ordering you choose, **it must be the same for both sets of data**.

In this example, let us rank the data from highest to lowest. If two or more values are the same, see the 'Tip' box on page 63 for how to rank them. It is best to do this step in a table; see the third and fifth columns of Table 5.4.

Table 5.4 Ranked data for GNI per capita and health expenditure per capita

Country	GNI per capita	Rank	Health expenditure per capita	Rank	d	d^2
Australia	65 000	3	6 000	3	0	0
Norway	104 000	1	10 000	1	0	0
Japan	40 000	4	4 000	4	0	0
Switzerland	88 000	2	9 000	2	0	0
Spain	29 000	5	3 000	5	0	0
Chile	15 000	6	1 000	6.5	−0.5	0.25
Mexico	10 000	8	700	8	0	0
Russia	13 000	7	1 000	6.5	0.5	0.25
Peru	6 000	11	400	10	1	1
China	7 000	9.5	400	10	−0.5	0.25
Botswana	7 000	9.5	400	10	−0.5	0.25
Bangladesh	1 000	13.5	30	15	−1.5	2.25
Ghana	2 000	12	100	12	0	0
Uganda	1 000	13.5	60	13	0.5	0.25
Nepal	700	15	40	14	1	1
					$\sum d^2 = 5.5$	

Step 3: for each pair of data values, calculate the difference, d, between the two ranks; this can be a positive or negative value. Make an extra column in the table to enter these differences (the sixth column in Table 5.4).

Step 4: square the d values to eliminate negative signs. Make an extra column in the table to enter these d^2 values (the last column in Table 5.4).

Step 5: add up all the d^2 values.

$$\sum d^2 = 5.5$$

Step 6: insert the values for $\sum d^2$ and n into the formula and calculate the R_s value.

$$R_s = 1 - \frac{6\sum d^2}{n^3 - n}$$

$$= 1 - \frac{6 \times 5.5}{15^3 - 15}$$

$$= 1 - \frac{33}{3375 - 15}$$

$$= 1 - \frac{33}{3360}$$

$$= 1 - 0.00982\ldots$$

$$= 1 - 0.99017\ldots \approx 0.99$$

Step 7: interpret the R_s value.

An R_s value of 0.99 indicates a strong positive correlation.

But is this result statistically significant? In other words, is it possible that it has occurred by chance?

Step 8: test the significance of the result using the Spearman rank test.

To apply the test you need the **degrees of freedom** (df) of the data and the **significance level** you want to test at.

■ The degrees of freedom for the Spearman rank test is the number of paired values in the data set, in this case 15.
■ The significance level most commonly used in geography is 0.05, which means that there is only a 5 in 100 likelihood of the result having occurred by chance, or that 95 out of 100 investigations would be expected to yield the same result. The 0.01 significance level is sometimes used.

A table of critical values for the Spearman rank test is given in Appendix 1.
Looking up the entry for 15 degrees of freedom at a significance level of 0.05, we see that the critical value is 0.441.

■ If the size of the R_s value you have calculated is less than the critical value, then accept the null hypothesis of no relationship.
■ If the R_s value is greater in size than the critical value, then reject the null hypothesis of no relationship. This means that there is a significant relationship between the two factors.

In this case $R_s = 0.99$ is well above 0.441. In fact it is also greater than 0.646, the critical value for 14 degrees of freedom and significance level 0.01. This means that there is less than a 1 in 100 chance of the result having occurred purely randomly.

Step 9: make a concluding statement.

Therefore, we can reject the null hypothesis and assert with a high degree of certainty (99%) that there is a strong positive correlation between GNI and health expenditure.

TIP

If you need to rank a set of values, and two of the values are the same, then the mean of their positions in the ordering is calculated and both percentages are assigned this value as their rank.

For example, in the ordered sequence 10, 9, 8, 8, 7, 6, …, the two '8's are in positions 3 and 4.

$$(3 + 4) \div 2 = 7 \div 2 = 3.5$$

so both '8's are assigned the rank 3.5 and the next value in the sequence (7) has rank 5.

The same principle applies if more than two values are the same. For example, in the ordered sequence 10, 9, 8, 7, 7, 7, 6, 5, …, the three '7's are in positions 4, 5 and 6.

$$(4 + 5 + 6) \div 3 = 15 \div 3 = 5$$

so all three '7's are assigned the rank 5 and the next value in the sequence (6) has rank 7.

The Spearman rank correlation coefficient test can also be used to examine how data changes over time. In this case, time is the independent variable which will be rank ordered.

B Guided question

Copy out the workings and complete the answers on a separate piece of paper.

1   **Use the data in Table 5.5 to perform a Spearman rank correlation test.**

Table 5.5 Average Arctic sea ice extent, 1996–2011

Year	Arctic sea ice extent (million km^2)
1996	9.4
1997	8.7
1998	8.8
1999	9.1
2000	8.9
2001	8.6
2002	8.7
2003	8.6
2004	8.5
2005	8.5
2006	8.3
2007	6.7
2008	8.4
2009	7.5
2010	7.7
2011	7.0

Source: National Snow and Ice Data Center, USA

a **Step 1: state the null hypothesis for the statistical test.**

There is _____ between time and Arctic sea ice extent.

b **Step 2: rank both sets of values in the data from lowest to highest.**

On a copy of Table 5.6, enter the rankings in the second and fourth columns.

Table 5.6 Ranked average Arctic sea ice extent data

Year	Rank	Arctic sea ice extent (million km²)	Rank	d	d^2
1996	1	9.4			
1997	2	8.7			
1998		8.8			
1999		9.1			
2000		8.9			
2001		8.6			
2002		8.7			
2003		8.6			
2004		8.5			
2005		8.5			
2006		8.3			
2007	12	6.7	1	11	121
2008		8.4			
2009		7.5			
2010		7.7			
2011	16	7.0	2	14	196
					$\sum d^2 =$

c **Steps 3 and 4: for each pair of data values, calculate the difference, d, between the two ranks. Then square the d values to eliminate negative signs.**

Enter these d and d^2 values in the last two columns of Table 5.6.

d **Step 5: add up all the d^2 values to find $\sum d^2$.**

Enter this in the last row of Table 5.6.

e **Step 6: insert the values for $\sum d^2$ and n into the formula and calculate the R_S value.**

$$R_S = 1 - \frac{6\sum d^2}{n^3 - n}$$

$$= 1 - \frac{6 \times \boxed{}}{16^3 - 16}$$

$$= 1 - \frac{\boxed{}}{\boxed{} - \boxed{}}$$

$$= \underline{}$$

f **Step 7: interpret the R_S value.**

The R_S value of _____ indicates _____ correlation.

g **Step 8: test the significance of the result at the 0.01 significance level.**

Degrees of freedom = 16

From the table for the Spearman rank test in Appendix 1, the critical value for this number of degrees of freedom and significance level 0.01 is _____

The size of the R_s value found in part **e** is greater/less than the critical value, so reject/accept the null hypothesis.

h **Step 9: make a concluding statement.**

There is _____ relationship between time and Arctic sea ice extent: as time increases, the Arctic sea ice extent tends to _____.

C Practice question

2 Analyse the relationship in the data given in Table 5.7.

  a Draw a scatter graph for the two variables.
  b Add a line of best fit.
  c Circle any anomalies.
  d Write a null hypothesis for the Spearman rank test.
  e Calculate and interpret the Spearman rank correlation coefficient.
  f Test the significance of the result.
  g Make a concluding statement.

Table 5.7 GNI per capita (in US dollars rounded to nearest thousand) and average number of deaths per flood event in selected countries

Country	GNI per capita	Average deaths per flood event
USA	55 000	9
UK	43 000	1
Spain	29 000	13
Italy	34 000	27
China	7 000	453
Canada	52 000	3
Austria	50 000	4
Algeria	6 000	91
Bangladesh	1 000	258
India	2 000	372
Egypt	4 000	154
Nepal	700	227
Japan	42 000	47
Pakistan	3 000	176
Mexico	9 000	50
Brazil	12 000	40

Sources: World Bank and Red Cross

The Spearman rank test is best used with at least 10 pairs of data values and it is limited to 30 pairs of data values.

When ranking the values, it does not matter whether you rank from high to low or from low to high, but whichever ordering you choose, it must be the same for both sets of data.

Geographical application

Fieldwork and the independent investigation

- In geography fieldwork, we often seek to study the relationship between factors. The Spearman rank correlation coefficient is a numerical value between −1 and 1 that describes the nature (positive, negative or none) and strength of correlation between two factors — such as between slope angle and soil depth, or between a country's wealth and the percentage of its population employed in agriculture. The Spearman rank test is used to determine whether or not the apparent relationship between two variables is statistically significant.

- However, even if the Spearman rank test indicates a significant strong correlation, this does not necessarily mean that there is a cause-and-effect relationship between the factors — it just tells us that some sort of correlation exists. In order to explain the result, further research and analysis based on geographical knowledge is necessary.

- In fieldwork reports and the independent investigation, much of the analysis about statistical tests will be in the form of written descriptions and explanations of the results. In this written work you can discuss possible cause-and-effect relationships among the factors.

Results interpretation

Table 5.8 suggests some questions that you could consider in order to interpret the result of the Spearman rank test.

Table 5.8

Geographical context of example	Questions leading to a possible explanation of the result
Wealth (GNI) and healthcare expenditure (**Table 5.1, page 57**)	How might increased wealth lead to more government spending on healthcare?
	Tax revenues are higher in wealthier countries — how could this influence spending on healthcare?
	How does personal income affect healthcare spending?
	How might expenditure on research and development in healthcare be affected?
Wealth and protection of coastal waters (**Table 5.2, page 59**)	What form does the coastal protection take — environmental protection, security, protection of resources?
	Look at the data for individual countries — are there some low-income countries with high levels of protection and why might this be?
Wealth and employment in agriculture (**Table 5.3, page 60**)	Think about the proportion of population employed in primary, secondary and tertiary industries in countries at different stages of development.
	Why are low-income countries more dependent on food production within their own country?
	What advantages does wealth have in terms of food production? (For example, consider levels of use of machinery versus manual labour.)
Changes in Arctic sea ice over time (**Table 5.5, page 63**)	What environmental and climatic factors could lead to increasing rates of reduction in sea ice? What is the evidence?
	How does the trend relate to previous patterns of change?

6 Testing for differences between data sets

In Chapter 5 you learned about the Spearman rank test, which can be used to decide whether there is significant **correlation** between two variables or data sets. In this chapter you will learn how to use three statistical tests for deciding whether or not there is a significant **difference** between two sets of values. The **Student's t test** and the **Mann–Whitney U test** can be applied to compare two sets of data, while the **chi-squared test** is used to compare observed data with 'theoretical' results expected from a hypothesis.

Student's t test

The Student's t test compares the **means** of two sets of data to decide if there is a significant difference between them. It will tell you whether the difference is conclusive, in a statistical sense, and to what extent it could have occurred just by chance. In geographical analysis, the next step would be to seek possible reasons for the difference.

For two data sets whose values are denoted by x and y, the formula for Student's t statistic is

$$t = \frac{|\bar{x} - \bar{y}|}{\sqrt{\dfrac{\dfrac{\Sigma x^2}{n_x} - \bar{x}^2}{n_x - 1} + \dfrac{\dfrac{\Sigma y^2}{n_y} - \bar{y}^2}{n_y - 1}}}$$

where:

- $\bar{x}$ and $\bar{y}$ are the means of the two data sets.
- Σx and Σy are the sums of each sample.
- n_x and n_y are the numbers of values in each data set.

Note that $|\bar{x} - \bar{y}|$ means the 'absolute difference' between $\bar{x}$ and $\bar{y}$; it is the **size** of the difference between $\bar{x}$ and $\bar{y}$ and is always positive.

As with all statistical tests, the Student's t test starts with a **null hypothesis**. This assumes that there is **no difference** between the two sets of data. The test will then establish whether or not this assumption can be **rejected** (meaning the data sets are significantly different).

Unlike the other tests covered in this chapter, the Student's t test is a **parametric test**. This means that it should only be used when the data follows a **normal distribution** (see Chapter 2, Figure 2.2).

A Worked example

As part of an ecosystem study at Sefton Dunes, Merseyside, students conducted a quadrat survey at two locations, A (20 m from the sea) and B (25 m from the sea). At each location there were ten sample sites and at each site the richness of species was represented by the total number of individual species found in a single quadrat. Based on the data in Table 6.1, use the Student's t test to determine if there is a significant difference between the numbers of plant species at locations A and B.

Table 6.1 Richness of species (number of individual species found in a quadrat) at two coastal locations on Sefton Dunes

Site number	Location A	Location B
1	3	3
2	6	6
3	13	13
4	8	15
5	8	13
6	6	14
7	9	11
8	5	7
9	11	15
10	15	16

Step 1: state the null hypothesis.

The richness of species is the same at location A as at location B.

Step 2: calculate the values needed for the formula.
These are $\bar{x}$, $\bar{y}$, $\sum x^2$ and $\sum y^2$. It is best to construct a table (see Table 6.2) to calculate them.

Let us take the data at location A to be the x values and the data at location B to be the y values. Then we know that $n_x = n_y = 10$.

To calculate the means:

$$\sum x = 3 + 6 + 13 + 8 + 8 + 6 + 9 + 5 + 11 + 15 = 84$$

so $\bar{x} = 84 \div 10 = 8.4$

$$\sum y = 3 + 6 + 13 + 15 + 13 + 14 + 11 + 7 + 15 + 16 = 113$$

so $\bar{y} = 113 \div 10 = 11.3$

Similarly, add up the values in the third column of Table 6.2 to find $\sum x^2$ and add up the values in the fifth column of Table 6.2 to find $\sum y^2$.

Table 6.2

Site number	Species richness at A (x values)	x^2	Species richness at B (y values)	y^2
1	3	9	3	9
2	6	36	6	36
3	13	169	13	169
4	8	64	15	225
5	8	64	13	169
6	6	36	14	196
7	9	81	11	121
8	5	25	7	49
9	11	121	15	225
10	15	225	16	256
Totals:	$\sum x = 84$	$\sum x^2 = 830$	$\sum y = 113$	$\sum y^2 = 1455$
	$\bar{x} = 8.4$		$\bar{y} = 11.3$	

The absolute difference $|\bar{x} - \bar{y}|$ is $|8.4 - 11.3| = 2.9$ (remember that this must be a positive value).

Step 3: insert the values into the t test formula and calculate the t value.

$$t = \frac{|\bar{x} - \bar{y}|}{\sqrt{\dfrac{\dfrac{\sum x^2}{n_x} - \bar{x}^2}{n_x - 1} + \dfrac{\dfrac{\sum y^2}{n_y} - \bar{y}^2}{n_y - 1}}}$$

$$= \frac{2.9}{\sqrt{\dfrac{\dfrac{830}{10} - 8.4^2}{10 - 1} + \dfrac{\dfrac{1455}{10} - 11.3^2}{10 - 1}}}$$

$$= \frac{2.9}{\sqrt{\dfrac{83 - 70.56}{9} + \dfrac{145 - 127.69}{9}}}$$

$$= \frac{2.9}{\sqrt{\dfrac{12.44}{9} + \dfrac{17.31}{9}}}$$

$$= \frac{2.9}{\sqrt{1.382 + 1.923}}$$

$$= \frac{2.9}{\sqrt{3.3}}$$

$$= \frac{2.9}{1.8}$$

$$= 1.6$$

Step 4: test the significance of the result using the table of critical values for Student's t test in Appendix 1.

For Student's t test the degrees of freedom is

$$(n_x - 1) + (n_y - 1) = 9 + 9 = 18$$

For 18 degrees of freedom, the critical value at the 0.05 significance level is 1.73 and the critical value at the 0.01 significance level is 2.88.

Similarly to the Spearman rank test:

■ If the t value you have calculated is less than the critical value, then accept the null hypothesis of no difference between the data sets.

■ If the t value is greater than the critical value, then reject the null hypothesis. This means that there is a significant difference between the two data sets.

In this case, the t value of 1.6 is less than both of the critical values, so we accept the null hypothesis.

Step 5: make a concluding statement.

The species richness at location A and at location B is not significantly different.

Despite not finding a significant difference, it is important to realise that the inquiry has still been worthwhile and is as valid a piece of research as finding a significant difference — a negative result does not mean that something must have been done incorrectly or inaccurately.

As always, in geography the next step would be to seek reasons for the result; in this case, it could be that the two sample sites are too close together to show any meaningful difference in species richness.

B Guided question

Copy out the workings and complete the answers on a separate piece of paper.

1 **Secondary data was collected on access to improved water sources in urban and rural areas in a sample of African countries. Use the Student's t test to determine if there is a significant difference between the urban and rural data sets.**

Table 6.3 Access to improved water sources in urban and rural areas of selected countries in Africa, 2015

Country	Urban areas % of population with access to improved water sources	Rural areas % of population with access to improved water sources
Tanzania	77	46
Kenya	82	57
Mozambique	81	37
Zimbabwe	97	67
Ethiopia	93	49
Uganda	96	76
Niger	100	49
Morocco	99	65
Senegal	93	67
Cameroon	95	53

Source: World Bank

Step 1: state the null hypothesis.

Urban areas have _____ level of access to improved water sources compared to rural areas.

Step 2: calculate the $\bar{x}$, $\bar{y}$, $\sum x^2$ and $\sum y^2$ values needed for the formula. Do this by filling in a copy of Table 6.4.

Take the data for urban areas to be the x values and the data for rural areas to be the y values. Then we know that $n_x = n_y =$ ____.

Table 6.4

Country	Urban areas (x values)	x^2	Rural areas (y values)	y^2
Tanzania	77	5 929	46	2 116
Kenya	82	6 724	57	
Mozambique	81		37	
Zimbabwe	97		67	
Ethiopia	93		49	
Uganda	96		76	
Niger	100		49	
Morocco	99		65	
Senegal	93		67	
Cameroon	95		53	
Totals:	$\sum x =$	$\sum x^2 =$	$\sum y =$	$\sum y^2 =$
	$\bar{x} =$		$\bar{y} =$	

The absolute difference $|\bar{x} - \bar{y}|$ is $|$___ $-$ ___$| =$ ____

Step 3: insert the values into the formula and calculate the t value.

$$t = \frac{|\bar{x} - \bar{y}|}{\sqrt{\dfrac{\dfrac{\sum x^2}{n_x} - \bar{x}^2}{n_x - 1} + \dfrac{\dfrac{\sum y^2}{n_y} - \bar{y}^2}{n_y - 1}}}$$

$$= \frac{\square}{\sqrt{\dfrac{\dfrac{}{10-1} - \left(\square\right)^2}{10-1} + \dfrac{\dfrac{}{10-1} - \left(\square\right)^2}{10-1}}}$$

$$= \frac{\square}{\sqrt{\dfrac{}{9} + \dfrac{}{9}}}$$

$$= \frac{\square}{\sqrt{}}$$

$$= \text{———}$$

$$= \text{———}$$

Step 4: test the significance of the result.

The degrees of freedom is $(n_x - 1) + (n_y - 1) =$ _____ $+$ _____ $=$ _____

From the table of critical values for Student's t test in Appendix 1, the critical value at the 0.01 significance level is _____.

The calculated t value _____ is less/greater than the critical value, so we accept/reject the null hypothesis.

Step 5: make a concluding statement.

There is _____ difference in the level of access to improved water sources in urban and rural areas.

C Practice question

2   As part of fieldwork related to decomposition in the carbon cycle, students collected soil samples from slopes with a north-facing and a south-facing aspect. The depth of the soil was also measured to study soil formation processes. The data is summarised in Table 6.5.

Use the Student's t test to determine if there is a significant difference between:
a   the depth
b   the percentage of organic matter

of the soil on north-facing and south-facing slopes.

For the test, take the north-facing data to be the x values and the south-facing data to be the y values.

Table 6.5 Depth and percentage organic matter of soil samples collected from north- and south-facing slopes

Sample	North-facing slope		South-facing slope	
	Soil depth (cm)	% organic matter	Soil depth (cm)	% organic matter
1	15	41	75	10
2	9	45	35	11
3	10	38	70	12
4	8	37	50	9
5	7	30	37	13
6	14	40	26	15
7	13	51	51	17
8	12	33	25	11

Mann–Whitney U test

Like the Student's t test, the Mann–Whitney U test is used to determine whether there is a significant **difference** between two data sets, but instead of comparing the means it compares the **median** values of the two samples.

The test is **non-parametric**, which means that it makes no assumption of the data being normally distributed. Because of this, it is a very flexible test with wide-ranging applications. It can also be applied to small data sets, having a number of values above 5 and no more than 20. The two data sets being tested need not have the same number of values, but all the values will be put in a single rank order (rather than being ranked separately as in the Spearman rank test).

The test starts with a **null hypothesis** that there is **no difference** between the two sets of data. This assumption will be **rejected** — meaning that there **is** a significant difference — if the calculated U value is **less than** a critical value, which can be looked up in the relevant table in Appendix 1. Note that this comparison of the calculated value and the critical value differs from the comparisons made in the Spearman rank test or Student's t test, where the null hypothesis is rejected if the calculated value is greater than the critical value.

The Mann–Whitney U test proceeds as follows:

- Step 1: state the null hypothesis.
- Step 2: take all the values of the two data sets and rank them in order, keeping track of which data set each value belongs to. If two or more values are the same, follow the instructions in the 'Tip' box in the Spearman rank test section (page 63).
- Step 3: calculate the U values for both data sets. The formulae are

$$U_x = n_x \times n_y + \frac{n_x(n_x+1)}{2} - \sum r_x$$

$$U_y = n_x \times n_y + \frac{n_y(n_y+1)}{2} - \sum r_y$$

where:

- n_x and n_y are the numbers of values in each data set.
- r_x and r_y are the ranks of the values in each data set.

- Step 4: test the significance of the result using the table of critical values for the Mann–Whitney U test in Appendix 1.

  The calculated value that should be compared with the critical value is the **smaller** of U_x and U_y.

- Step 5: make a concluding statement.

A Worked example

The secondary data in Table 6.6 represents the air pollution levels in a selection of cities in advanced countries (ACs) and emerging and developing countries (EDCs). Is there a significant difference between the air quality in the two sets of cities?

Table 6.6 Air quality index for a selection of cities in ACs and EDCs, 2015 (values of index: 0–50, good; 51–100, moderate; 101–200, unhealthy; 201–300, very unhealthy; 301–500, hazardous)

AC city	Air quality index	EDC city	Air quality index
Vancouver	60	Shanghai	175
Phoenix	57	Tianjin	104
Athens	109	Mumbai	117
London	78	Hyderabad	147
Rome	61	Santiago	145
Madrid	109	Quito	151
Budapest	89	Bogota	185
Reykjavik	32	Jakarta	90

Source: World Air Quality Index

Step 1: state the null hypothesis.

Cities in ACs have the same air quality as cities in EDCs.

Step 2: take all the values of the two data sets and rank them from highest to lowest.

Let us take the data for ACs to be the x values and the data for EDCs to be the y values. Then we know that $n_x = n_y = 8$.

The ranking is done in Table 6.7.

Table 6.7 Ranked air quality index values for a selection of cities in ACs and EDCs

AC city	AC index (x values)	Rank (r_x)	EDC city	EDC index (y values)	Rank (r_y)
Vancouver	60	14	Shanghai	175	2
Phoenix	57	15	Tianjin	104	9
Athens	109	7.5	Mumbai	117	6
London	78	12	Hyderabad	147	4
Rome	61	13	Santiago	145	5
Madrid	109	7.5	Quito	151	3
Budapest	89	11	Bogota	185	1
Reykjavik	32	16	Jakarta	90	10
		$\sum r_x = 96$			$\sum r_y = 40$

Step 3: calculate the U values for both data sets.

First add together the rank values separately for x and y:

$$\sum r_x = 14 + 15 + 7.5 + 12 + 13 + 7.5 + 11 + 16 = 96$$

$$\sum r_y = 2 + 9 + 6 + 4 + 5 + 3 + 1 + 10 = 40$$

Then substitute the values of n_x, n_y, $\sum r_x$ and $\sum r_y$ into the formulae:

$$U_x = n_x \times n_y + \frac{n_x(n_x + 1)}{2} - \sum r_x$$

$$= 8 \times 8 + \frac{8 \times (8 + 1)}{2} - 96$$

$$= 64 + \frac{72}{2} - 96$$

$$= 64 + 36 - 96$$

$$= 4$$

$$U_y = n_x \times n_y + \frac{n_y(n_y + 1)}{2} - \sum r_y$$

$$= 8 \times 8 + \frac{8 \times (8 + 1)}{2} - 40$$

$$= 64 + 36 - 96$$

$$= 60$$

Note: the sum $U_x + U_y$ should equal the product $n_x \times n_y$; if they don't match, then you have made a mistake. In this case $4 + 60 = 64 = 8 \times 8$.

Step 4: test the significance of the result using the table of critical values for the Mann–Whitney U test in Appendix 1.

The critical value is found in row n_x and column n_y of the table, so for this example we look at the entry in the 8th row and 8th column, which is 13.

The **smaller** of U_x and U_y is the value to compare with the critical value.
So we compare 4 with the critical value 13:

> 4 is **less than** 13

> so the null hypothesis should be **rejected**.

Step 5: make a concluding statement.

There is a significant difference in the air quality of cities in ACs and cities in EDCs.

B Guided question

Copy out the workings and complete the answers on a separate piece of paper.

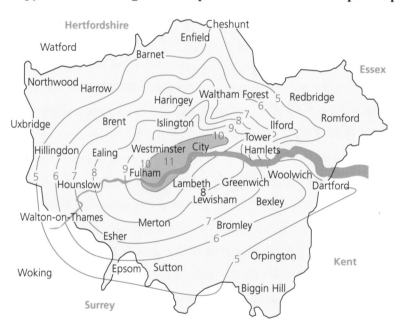

Figure 6.1 Isoline map of London 'heat island', example of an urban micro-climate

1 **Following a study of urban micro-climates in class (see figure 6.1), students collected some primary data to compare temperatures in urban and rural locations. Over a period of ten days in early May, a single temperature reading was taken at the same time each day in the middle of an urban area and at a location in the rural periphery. The results are summarised in Table 6.8. Use the Mann–Whitney U test to determine if there is a significant difference between temperatures at the urban and rural locations.**

Table 6.8 Air temperatures at an urban location and a rural location in early May

Reading	Urban location air temperature (°C)	Rural location air temperature (°C)
1	12	9
2	10.5	8
3	10	8
4	11	9
5	9.5	7
6	11.5	9
7	12	10
8	11	8
9	12.5	11
10	12.5	10

Step 1: state the null hypothesis.

The air temperature at the urban location is _____ the air temperature at the rural location.

Step 2: take all the values of the two data sets and rank them from highest to lowest.

Let us take the data for the urban location to be the x values and the data for the rural location to be the y values. Then we know that $n_x = n_y = $ _____

Complete the ranking in a copy of Table 6.9. Notice that there are many repeated values among the temperature readings, so you will need to assign ranks according to the 'Tip' box in the Spearman rank test section (page 63).

Table 6.9 Ranked air temperatures at an urban location and a rural location

Reading	Urban location temperature (x values)	Rank (r_x)	Rural location temperature (y values)	Rank (r_y)
1	12		9	
2	10.5		8	
3	10		8	
4	11		9	
5	9.5		7	20
6	11.5		9	
7	12		10	
8	11		8	
9	12.5	1.5	11	
10	12.5	1.5	10	
		$\Sigma r_x = $ _____		$\Sigma r_y = $ _____

Step 3: calculate the U values for both data sets.

Substitute the values of n_x, n_y, $\sum r_x$ and $\sum r_y$ into the formulae:

$$U_x = n_x \times n_y + \frac{n_x(n_x+1)}{2} - \sum r_x$$

$$= \_\_ \times \_\_ + \frac{\_\_ \times (\_\_+1)}{2} - \_\_\_\_$$

$$= \_\_\_\_ + \frac{\Box}{2} - \_\_\_\_$$

$$= \_\_\_\_$$

$$U_y = n_x \times n_y + \frac{n_y(n_y+1)}{2} - \sum r_y$$

$$= \_\_ \times \_\_ + \frac{\_\_ \times (\_\_+1)}{2} - \_\_\_\_$$

$$= \_\_\_\_ + \frac{\Box}{2} - \_\_\_\_$$

$$= \_\_\_\_$$

Remember to check your calculations: the sum $U_x + U_y$ should equal the product $n_x \times n_y$.

Step 4: test the significance of the result using the table of critical values for the Mann–Whitney U test in Appendix 1.

The critical value is the entry in row ___ and column ___ of the table, which is _____

The smaller value of U_x and U_y is _____

Compare this U value with the critical value and draw a conclusion.

- If the smaller U value is **less than** the critical value then the null hypothesis should be **rejected**.

- If the smaller U value is **more than** the critical value then the null hypothesis should be **accepted**.

Step 5: write a concluding statement.

There is a/no significant difference in the air temperature at the urban location and at the rural location.

C Practice question

2 As part of a fieldwork study to examine the impact of a local factory on the quality of river water, students collected seven water samples from each of three different locations:

- site 1 — 1 km upstream of the factory
- site 2 — 1 km downstream of the factory
- site 3 — 5 km downstream of the factory

From the water samples, water clarity, temperature, pH and dissolved oxygen content (DO) were measured. The results for DO are shown in Table 6.10.

Table 6.10 Dissolved oxygen content (DO) measured at three sites

Sample	Site 1 DO (mg/l)	Site 2 DO (mg/l)	Site 3 DO (mg/l)
1	8.5	3.5	6.0
2	6.5	4.0	5.5
3	7.0	3.0	6.0
4	7.0	2.5	7.0
5	6.0	4.5	5.0
6	7.5	3.5	6.5
7	8.0	2.5	7.0

a   Use the Mann–Whitney U test to determine if there is any difference between the DO at site 1 and at site 2.

b   Use the Mann–Whitney U test to determine if there is any difference between the DO at site 1 and at site 3.

c   What do the results in parts **a** and **b** tell you about the quality of river water in relation to the factory?

Chi-squared test

The chi-squared test is a **comparative** test that is used to compare **observed** and **expected** distributions and determine whether or not there is a significant difference between them. In most cases the 'expected' results are the 'theoretical' outcomes that would have arisen by chance.

For the chi-squared test to be used:
- The data needs to be organised into **categories**.
- The data values are the **frequencies** of the categories (they should not be percentages or proportions).
- The total number of observed data values must exceed 20.
- The **expected frequency** of each category should exceed 4.

The formula for the chi-squared value is

$$\chi^2 = \sum \frac{(O-E)^2}{E}$$

where:
- O is the observed data value.
- E is the expected frequency.
- Σ (Sigma, the Greek capital letter 'S') stands for 'sum'.

The steps of performing a chi-squared test are as follows.

- Step 1: state the null hypothesis, which assumes that there is no difference between the observed data and the expected results.
- Step 2: calculate the expected frequency of each category.

- Step 3: for each category, calculate the difference between the observed and expected frequencies, $O - E$, then square the difference to obtain $(O - E)^2$ and finally divide by the expected frequency to get $\frac{(O - E)^2}{E}$. These calculations are best organised in a table.
- Step 4: add up all the $\frac{(O - E)^2}{E}$ values to obtain the χ^2 value.
- Step 5: test the significance of the result using the table of critical values for the chi-squared test in Appendix 1.
  - □ If the calculated χ^2 value is **lower** than the critical value, then **accept** the null hypothesis.
  - □ If the calculated χ^2 value is **higher** than the critical value, then **reject** the null hypothesis.
- Step 6: make a concluding statement.

 A Worked example

As part of a farm study, soil samples were collected from different land-use areas and their water content was measured.

The four categories of land use were:
- **arable fields with sandy soil**
- **arable fields with clay soil**
- **grazing land**
- **deciduous woodland**

Of the soil samples taken, 44 had a water content exceeding 30%.

The results are summarised in Table 6.11.

Table 6.11 Soil samples with above 30% moisture content taken from four categories of land use

	Arable–sandy	Arable–clay	Grazing	Deciduous woodland	Total
Number of observed samples with moisture content >30%	5	9	10	20	44

Use the chi-squared test to determine whether the distribution of soil samples with above 30% moisture content is random.

Notice that the data consists of frequencies — the **number** of samples in each category. All of the frequencies are greater than 4 and the total is greater than 20, so the chi-squared test can be applied.

Step 1: state the null hypothesis.

The distribution of soil samples with above 30% moisture content is random.

Step 2: calculate the expected frequency of each category.

The expected frequencies are what you would expect to get if the distribution was completely random (that is, if the null hypothesis were true).

With a random distribution, we would expect a quarter of the 44 samples to fall into each of the four categories, so

$E = 44 \div 4 = 11$ for each category.

Appending the expected frequencies to Table 6.11, we get Table 6.12.

Table 6.12 Observed and expected frequencies of soil samples with above-30% moisture content taken from four categories of land use

	Arable–sandy	Arable–clay	Grazing	Deciduous woodland	Total
Observed frequency	5	9	10	20	44
Expected frequency	11	11	11	11	44

Step 3: for each category, calculate the difference between the observed and expected frequencies, $O - E$, then square the difference to obtain $(O - E)^2$ and finally divide by the expected frequency to get $\frac{(O - E)^2}{E}$. These calculations are done in Table 6.13.

Table 6.13

Land use category	Observed frequency (O)	Expected frequency (E)	$O - E$	$(O - E)^2$	$\dfrac{(O - E)^2}{E}$
Arable–sandy	5	11	−6	36	3.27
Arable–clay	9	11	−2	4	0.36
Grazing	10	11	−1	1	0.09
Deciduous woodland	20	11	9	81	7.36
					$\chi^2 = 11.08$

Step 4: add up all the $\frac{(O - E)^2}{E}$ values to obtain the χ^2 value.

$\chi^2 = 3.27 + 0.36 + 0.09 + 7.36 = 11.08$

Step 5: test the significance of the result.

Degrees of freedom = the number of categories $- 1 = 4 - 1 = 3$

From the table for the chi-squared test in Appendix 1, the critical value for 3 degrees of freedom is

- 7.82 at the 0.05 significance level
- 11.34 at the 0.01 significance level

The χ^2 value of 11.08 is greater than 7.82, so the result is significant at the 0.05 level.

This means that the null hypothesis can be rejected with 95% certainty.

Step 6: make a concluding statement.

The different frequencies of soil samples with above 30% moisture content from different categories of land use did not occur purely by chance.
There is an association between the type of land use and the soil moisture content.

B Guided question

Copy out the workings and complete the answers on a separate piece of paper.

1 As part of an urban study, students collected data on the number of fast food retail outlets with increasing distance from the central business district (CBD). A transect was marked on a base map covering the following categories of distance from the CBD (rounded to the nearest km):

- 0–1 km
- 2–3 km
- 4–5 km
- more than 5 km

The results are summarised in Table 6.14.

Table 6.14 Number of fast food outlets with increasing distance from the CBD

	0–1 km	2–3 km	4–5 km	>5 km	Total
Number of fast food retail outlets	16	8	3	1	28

Use the chi-squared test to decide whether the number of fast food outlets varies with increasing distance from the CBD.

Step 1: state the null hypothesis.

There is _____ in the number of fast food outlets with distance from the CBD.

Step 2: calculate the expected frequency of each category and enter these frequencies in a copy of Table 6.15.

Divide the total number of data values by the number of categories.

Table 6.15 Observed and expected numbers of fast food outlets with increasing distance from the CBD

	0–1 km	2–3 km	4–5 km	>5 km	Total
Observed frequency	16	8	3	1	28
Expected frequency					28

Step 3: for each category, calculate the difference between the observed and expected frequencies, $O - E$, then square the difference to obtain $(O - E)^2$ and finally divide by the expected frequency to get $\frac{(O - E)^2}{E}$. Enter these calculations in a copy of Table 6.16.

Table 6.16

Distance from CBD	Observed frequency (O)	Expected frequency (E)	$O - E$	$(O - E)^2$	$\frac{(O - E)^2}{E}$
0–1 km	16				
2–3 km	8				
4–5 km	3				
>5 km	1				

Step 4: add up all the $\frac{(O-E)^2}{E}$ values to obtain the χ^2 value.

$\chi^2 = \underline{\quad} + \underline{\quad} + \underline{\quad} + \underline{\quad} = \underline{\quad}$

Step 5: test the significance of the result.

Degrees of freedom = number of categories − 1 = $\underline{\quad}$ − 1 = $\underline{\quad}$

From the table for the chi-squared test in Appendix 1, the critical value is

- $\underline{\quad}$ at the 0.05 significance level
- $\underline{\quad}$ at the 0.01 significance level

The χ^2 value $\underline{\quad}$ is greater/less than the critical value $\underline{\quad}$, so the null hypothesis can be rejected/accepted with $\underline{\quad}$% certainty.

Step 6: make a concluding statement.

There is a/no significant difference in the number of fast food outlets with increasing distance from the CBD.

C Practice question

2   As part of a fieldwork activity on the perception of place, students conducted a survey using a **bipolar semantic scale** like the one in Figure 6.2.

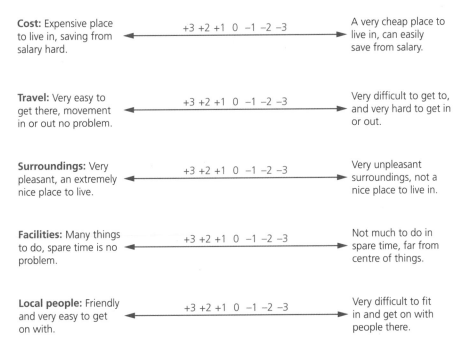

Figure 6.2 Bipolar semantic differential scale used to measure perception of place

Fifty people were asked to reflect on their perception of the local town centre. The same scale was used when respondents were shown a photograph of a 'scenic rural setting'. The results were sorted into categories based on the gender and approximate age group (younger than 25 or 25 and older) of the respondents; these are presented in Table 6.17.

Table 6.17 Results of place perception questionnaire

Age and gender category	Town centre			Rural 'idyll'		
	Number of respondents with a positive score	Number of respondents with a negative score	Row total	Number of respondents with a positive score	Number of respondents with a negative score	Row total
Female <25	10	4	14	5	14	19
Male <25	8	3	11	3	15	18
Female ≥25	5	8	13	9	0	9
Male ≥25	5	7	12	4	0	4
Column total	28	22	50	21	29	50

a   Use the chi-squared test to determine if there is any significant difference between the age/gender groups in positive perception of the town centre.

b   Use the chi-squared test to determine if there is any significant difference between the age/gender groups in negative perception of the rural 'idyll'.

TIP

When the categories of data are organised in a table with more than one row and more than one column, the expected frequencies can be calculated using the formula

$$\text{expected frequency} = \frac{\text{row total} \times \text{column total}}{\text{grand total}}$$

For example, in the Practice question, the expected number of respondents in the 'Female <25' group giving a positive score for the town centre is

$$\frac{\text{row total} \times \text{column total}}{\text{grand total}} = \frac{14 \times 28}{50} = 7.8$$

Some more of these calculations have been done for you in Table 6.18.

Table 6.18 Expected frequencies for the place perception example

	Town centre		Rural 'idyll'	
	Positive	Negative	Positive	Negative
Female <25	14 × 28 ÷ 50 = 7.8	14 × 22 ÷ 50 = 6.2		19 × 29 ÷ 50 = 11.0
Male <25			18 × 21 ÷ 50 = 7.6	
Female ≥25		13 × 22 ÷ 50 = 5.7		
Male ≥25	12 × 28 ÷ 50 = 6.7		4 × 21 ÷ 50 = 1.7	

Geographical application

Fieldwork and the independent investigation

- The three statistical tests covered in this chapter have a wide range of uses in fieldwork in both physical and human geography. The flow chart in Appendix 2 guides you through a decision-making process to select the most appropriate test given the aims of the study and the kind of data collected (number of samples, range and distribution of the data).

- Always bear in mind that once you have calculated a value for a statistical test, the result must be interpreted and then explained by drawing on your geographical understanding. For example, in the application of the Mann–Whitney U test to air quality readings in ACs and EDCs (Table 6.7), why does the air quality tend to be poor in cities of EDCs? How might this relate to rising incomes and rapid industrialisation? What other factors, such as government legislation, could be relevant?

- In explaining your results, you should also reflect on the data collection process and methodology. For example, after applying the Student's t test to the data on coastal species richness (Table 6.1), consider whether a random sample was the best way of collecting the data. Could the **sampling** method have affected the result? Were the sampling locations A and B far enough apart?

- Similarly, upon using the chi-squared test to analyse the data on place perception in Table 6.17, think about the possible disadvantages of using a subjective data collection method such as a bipolar semantic scale. How might it have affected the results?

Uses of statistical tests in geography

In many textbooks, statistical techniques are integrated into the theoretical content and are used to complement and enhance geographical understanding.

Table 6.19 suggests some links to geographical understanding.

Table 6.19

Geographical context of example	Questions leading to a possible explanation of the result
Richness of species at coastal locations (**Table 6.1, page 69**)	How do these results relate to theories of ecological succession in sand dune environments?
Organic content of soil samples (**Table 6.5, page 73**)	What are the factors affecting organic content of soils, e.g. temperature, vegetation, bacterial activity?
Air quality index values in ACs and EDCs (**Table 6.7, page 75**)	Analysis of patterns of air pollution relate to theories within topics such as sense of place, health, pollution, climate change and urbanisation. In particular, what is the role of air pollution in the debate on climate change? How can rising levels of pollution in urban areas be controlled and managed?
Air temperature at urban and rural locations (**Table 6.9, page 77**)	What factors lead to the intensity of the urban heat island effect, e.g. pollution, anthropogenic heating?
Place perception (**Table 6.17, page 84**)	How is a sense of place and an individual's perception of place influenced by demographic factors such as age and gender?

AS level questions

1 Figure E.1 plots the two variables beach width and average beach slope angle.

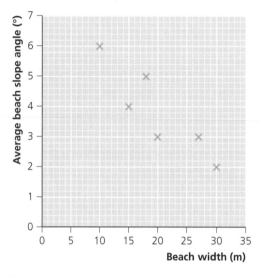

Figure E.1

    a   **i** Identify the independent variable in Figure E.1. **(1)**

        ii Add a line of best fit to Figure E.1 and state the nature of the relationship
between the two variables. **(2)**

    b Suggest **one** reason why the beach slope angle might increase as beach width
decreases. **(3)**

2 Table E.1 shows data on the distribution of wealth in the USA and Belgium.

Table E.1 Wealth distribution in the USA and Belgium

| Cumulative % of population | USA | | Belgium | |
	% wealth held by each 10% of population from poor to rich	Cumulative % of wealth	% wealth held by each 10% of population from poor to rich	Cumulative % of wealth
(Poorest 10%) 10	0.5	0.5	3	3
20	0.8	1.3	4	7
30	1.2	2.5	5	12
40	2.5	5.0	7	19
50	3.0	8.0	8	27
60	4.2	12.2	10	37
70	4.8	17.0	12	49
80	7.0	24.0	15	64
90	25.6	49.6	16	80
100 (Richest 10%)	50.4	100	20	100

The Lorenz curve for Belgium is plotted in Figure E.2.

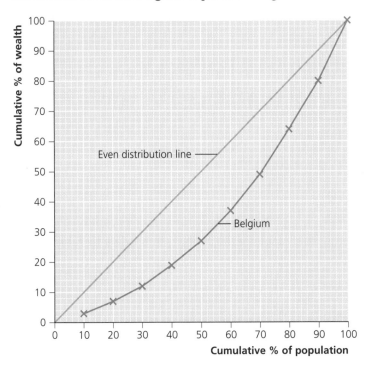

Figure E.2

a i Using the data in Table E.1, add a Lorenz curve for the USA to Figure E.2. **(2)**

ii Which country has the more equitable wealth distribution? **(1)**

b Suggest **one** cause of income inequality within a country. **(3)**

3 Table E.2 shows the top ten refugee-hosting countries in 2005 and 2014, together with their GNI per capita.

Table E.2 Top ten refugee-hosting countries and their GNI per capita, 2005 and 2014

	2005			2014	
Country	GNI per capita (US$)	Number of refugees	Country	GNI per capita (US$)	Number of refugees
Pakistan	1 260	1 084 700	Turkey	10 830	1 600 000
Iran	7 010	716 400	Pakistan	1 400	1 500 000
Germany	46 390	700 000	Lebanon	10 030	1 150 000
Tanzania	780	548 800	Iran	7 220	982 000
USA	52 536	379 300	Ethiopia	550	659 500
China	5 870	301 000	Jordan	5 160	654 100
UK	41 010	293 500	Kenya	1 290	551 400
Chad	960	275 400	Chad	980	452 900
Uganda	630	257 300	Uganda	670	385 500
Kenya	1 090	251 271	Chile	14 910	301 000

Source: United Nations High Commissioner for Refugees (UNHCR)

a i Sketch **one** type of suitable graphical presentation for the data in Table E.2. **(3)**

ii Suggest how the data in Table E.2 could be analysed using an appropriate quantitative technique. **(3)**

b Explain two push factors for refugees. **(6)**

4   As part of a study into coastal processes, a student collected pebble samples at two different beaches. The results from beach A are recorded in Table E.3.

Table E.3 Pebble sample, beach A

Sample number	1	2	3	4	5	6	7	8	9	10	11
A–axis measurement in cm	2.5	3.1	4.6	2.2	1.9	4.8	5.2	6.2	3.3	4.2	4.7

Upper auartile (UQ) $= \dfrac{n+1}{4}$

Lower quartile (LQ) $= \dfrac{n+1}{4} \times 3$

Interquartile range = UQ − LQ

n = number in sample

Calculate the mean and interquartile range for the pebble sample data in Table E.3 and interpret your findings.   **(6)**

A-level questions

1   Table E.4 shows calculations for finding the Spearman rank correlation coefficient for the number of internet users (per 100 population) and the level of exports of goods and services (as a percentage of GDP) in various countries.

Table E.4 Ranked data for number of internet users and exports of goods and services

| Country | Internet users | | Exports | | Rank difference | |
	Internet users per 100 people	Rank	Exports of goods and services (% GDP)	Rank	d	d^2
Afghanistan	6.4	12	6.6	14	−2	4
Mexico	44.4	8	32.4	7	1	1
Botswana	**18.5**	**10**	**49.8**			
China	49.3	6	22.6	12	−6	36
Greece	63.2	4	32.7	6	−2	4
Burundi	1.4	14	7.8	13	1	1
Liberia	5.4	13	25.0	10	3	9
Bolivia	39.0	9	43.3	4	5	25
Norway	96.3	1	38.3	5	−4	16
Turkey	51.0	5	27.7	9	−4	16
UAE	90.4	3	98.0	1	2	4
UK	91.6	2	28.4	8	−6	36
Vietnam	**48.3**	**7**	**86.4**			
India	18.0	11	23.2	11	0	0
					$\sum d^2 =$	

Source: World Bank

a i Fill in the missing values for Botswana and Vietnam. Then calculate the
       Spearman rank correlation coefficient (R_s) using the following formula. **(4)**

$$R_s = 1 - \frac{6\sum d^2}{n^3 - n}$$

   ii Use the critical values in Table E.5 to interpret the result of **i** and make a
       concluding comment referring to the following null hypothesis: **(3)**

**There is no significant relationship between the levels of internet use and
exports of goods and services.**

Table E.5 Extract of Spearman rank test critical values

Degrees of freedom	Significance (confidence) level	
	0.05 (95%)	0.01 (99%)
13	0.475	0.673
14	0.457	0.646
15	0.441	0.623

b Explain how internet usage can affect a country's ability to trade. **(6)**

2 Table E.6 shows calculations for finding the chi-squared value to determine whether
   the number of infiltration tests needed to drain 5 cm of water in less than half an hour
   is affected by location on the slope.

Table E.6 Chi-squared value calculations for the number of infiltration tests (out of
20) taking less than 30 minutes to drain 5 cm of water on a slope

Position on slope	Observed frequency (O)	Expected frequency (E)	$O - E$	$(O - E)^2$	$\frac{(O - E)^2}{E}$
Top	18	12	6	36	
Middle	15	12	3	9	
Base	3	12	−9	81	

a i Complete the $\frac{(O-E)^2}{E}$ column and calculate the chi-squared value $\chi^2 = \sum \frac{(O-E)^2}{E}$. **(2)**

   ii Use the critical values in Table E.7 to interpret the result of **i** and make a
       concluding comment referring to the following null hypothesis: **(3)**

**There is no significant difference in the infiltration rate at different points
on a slope profile.**

Table E.7 Selection of chi-squared critical values

Degrees of freedom	Significance (confidence) level	
	0.05 (95%)	0.01 (99%)
2	5.99	9.21
3	7.82	11.34
4	9.49	13.28

b Explain the factors which could influence infiltration rates. **(6)**

Appendix 1

Statistical tables

Critical values of Spearman rank correlation test

Degrees of freedom	Significance level	
	0.05	0.01
4	1.000	
5	0.900	1.000
6	0.829	0.943
7	0.714	0.893
8	0.643	0.833
9	0.600	0.783
10	0.564	0.745
11	0.523	0.736
12	0.497	0.703
13	0.475	0.673
14	0.457	0.646
15	0.441	0.623
16	0.425	0.601
17	0.412	0.582
18	0.399	0.564
19	0.388	0.549
20	0.377	0.534
21	0.368	0.521
22	0.359	0.508
23	0.351	0.496
24	0.343	0.485
25	0.336	0.475
26	0.329	0.465
27	0.323	0.456
28	0.317	0.448
29	0.311	0.440
30	0.305	0.432

Degrees of freedom = the number of paired values

Reject the null hypothesis if the calculated value is greater than the critical value at the chosen confidence limit.

Critical values of Student's t test

Degrees of freedom	Significance level	
	0.05	0.01
1	6.31	63.66
2	2.92	9.93
3	2.35	5.84
4	2.13	4.60
5	2.00	4.03
6	1.94	3.71
7	1.89	3.50
8	1.86	3.36
9	1.83	3.25
10	1.81	3.17
11	1.80	3.11
12	1.78	3.06
13	1.77	3.01
14	1.76	2.98
15	1.75	2.95
16	1.75	2.92
17	1.74	2.90
18	1.73	2.88
19	1.73	2.86
20	1.73	2.85
21	1.72	2.83
22	1.72	2.82
23	1.71	2.81
24	1.71	2.80
25	1.71	2.79
26	1.71	2.78
27	1.70	2.77
28	1.70	2.76
29	1.70	2.76
30	1.70	2.75

Degrees of freedom = where the two sample sizes are A and B, $(A - 1) + (B - 1)$

Reject the null hypothesis if the calculated value is greater than the critical value at the chosen confidence limit.

Critical values of chi-squared test

Degrees of freedom	Significance level	
	0.05	0.01
1	3.84	6.64
2	5.99	9.21
3	7.82	11.34
4	9.49	13.28
5	11.08	15.09
6	12.59	16.81
7	14.07	18.48
8	15.51	20.09
9	16.92	21.67
10	18.31	23.21
11	19.68	24.72
12	21.03	26.22
13	22.36	27.69
14	23.68	29.14
15	25.00	30.58
16	26.30	32.00
17	27.59	33.41
18	28.87	34.80
19	30.14	36.19
20	37.57	37.57
21	32.67	38.93
22	33.92	40.29
23	35.18	41.64
24	36.43	42.98
25	37.65	44.31
26	35.88	45.64
27	40.11	46.96
28	41.34	48.28
29	42.56	45.59
30	43.77	50.89
40	55.76	63.69
50	67.51	76.15
60	79.08	88.38
70	90.53	100.43
80	101.88	112.33
90	113.15	124.12
100	124.34	135.81

Degrees of freedom = $n - 1$ (where n is the number of values)

Reject the null hypothesis if the calculated value is greater than the critical value at the chosen confidence limit.

Mann–Whitney *U* test

Critical values at 0.05 significance level

		n_y																		
n_x	1	2	3	4	5	6	7	8	9	10	11	12	13	14	15	16	17	18	19	20
1	–	–	–	–	–	–	–	–	–	–	–	–	–	–	–	–	–	–	–	–
2	–	–	–	–	–	–	–	0	0	0	0	1	1	1	1	1	2	2	2	2
3	–	–	–	–	0	1	1	2	2	3	3	4	4	5	5	6	6	7	7	8
4	–	–	–	0	1	2	3	4	4	5	6	7	8	9	10	11	11	12	13	13
5	–	–	0	1	2	3	5	6	7	8	9	11	12	13	14	15	17	18	19	20
6	–	–	1	2	3	5	6	8	10	11	13	14	16	17	19	21	22	24	25	27
7	–	–	1	3	5	6	8	10	12	14	16	18	20	22	24	26	28	30	32	34
8	–	0	2	4	6	8	10	13	15	17	19	22	24	26	29	31	34	36	38	41
9	–	0	2	4	7	10	12	15	17	20	23	26	28	31	34	37	39	42	45	48
10	–	0	3	5	8	11	14	17	20	23	26	29	33	36	39	42	45	48	52	55
11	–	0	3	6	9	13	16	19	23	26	30	33	37	40	44	47	51	55	58	62
12	–	1	4	7	11	14	18	22	26	29	33	37	41	45	49	53	61	61	65	69
13	–	1	4	8	12	16	20	24	28	33	37	41	45	50	54	59	67	67	72	76
14	–	1	5	9	13	17	22	26	31	36	40	45	50	55	59	64	74	74	78	83
15	–	1	5	10	14	19	24	29	34	39	44	49	54	59	64	70	80	80	85	90
16	–	1	6	11	15	21	26	31	37	42	47	53	59	64	70	75	86	86	92	98
17	–	2	6	11	17	22	28	34	39	45	51	57	63	67	75	81	93	93	99	105
18	–	2	7	12	18	24	30	36	42	48	55	61	67	74	80	86	99	99	106	112
19	–	2	7	13	19	25	32	38	45	52	58	65	72	78	85	92	106	106	113	119
20	–	2	8	13	20	27	34	41	48	55	62	69	76	83	90	98	112	112	119	127

n_x and n_y in the table represent the number of x and y values.

Take the smaller value/result from Ux and Uy as the calculated value.

Reject the null hypothesis if the calculated value is equal to or less than the critical value.

Appendix 2

Choosing statistical methods for geographical fieldwork

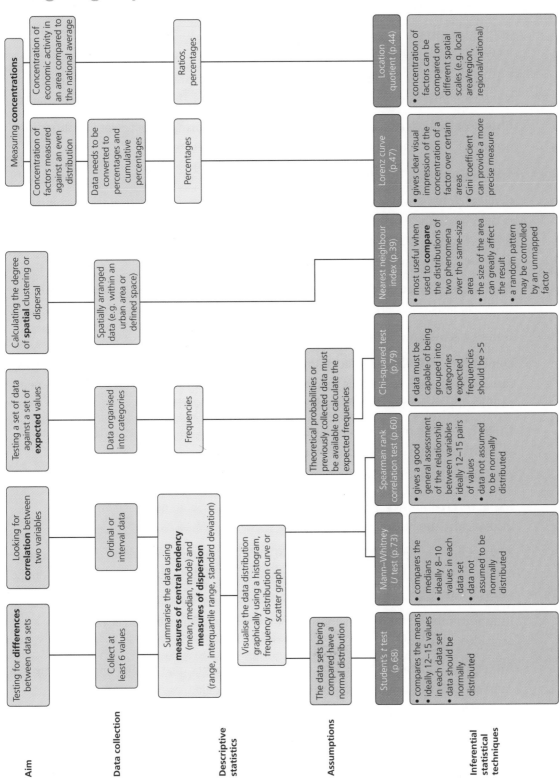

Key terms

Anomaly (residual) A data value that does not fit the general or expected trend in the data. When plotted on a scatter graph, anomalies lie far away from the line of best fit.

Central tendency A representative or 'average' value in a set of data. The mean, median and mode are measures of central tendency.

Chi-squared test A comparative statistical test to determine whether or not there is a significant difference between the observed data and the theoretically expected results.

Choropleth map A map with graded shading or colouring to represent variations in density of a characteristic.

Correlation The strength of the relationship between two variables.

Critical value A value with which the calculated result of a statistical test is compared in order to decide whether the null hypothesis should be accepted or rejected. Critical values represent the boundary between acceptance or rejection of the null hypothesis. They are usually presented in statistical tables.

Degrees of freedom A number representing the size of a sample, which is used in statistical tests. The definition can vary depending on the statistical test to be applied.

Density A measure of the concentration of a characteristic within a given area.

Dependent variable The factor that you measure in a study and which you expect to respond to changes in another variable (the independent variable). It is called 'dependent' because it is thought to be affected by the independent variable.

Dispersion The spread of the values in a data set. The range, interquartile range and standard deviation are measures of dispersion.

Distribution The spread of data across an area or over a collection of categories.

Expected frequency The results that you would expect to get based on some theoretical assumptions.

Histogram A graphical representation of the distribution of numerical data. The classes of data values are shown along the horizontal axis and the frequencies are plotted on the vertical axis.

Independent variable A factor that is expected to influence another factor (the dependent variable).

Interquartile range The difference between the upper quartile and lower quartile in a data set. It measures the dispersion or spread of the data.

Mann–Whitney U test A non-parametric statistical test that is used to compare the medians of two data sets and determine whether or not there is a significant difference between them.

Mean A commonly used measure of central tendency, or average, of a data set. All the data values are added together and then the total is divided by the number of values in the data set.

Median The middle value in a set of data which has been arranged in order. It is a measure of central tendency for the data set.

Mode The most frequently occurring value in a data set.

Non-parametric test A statistical test which does not assume that the data follows a normal distribution.

Normal distribution A distribution of data that is perfectly symmetrical and bell-shaped. On a normal distribution curve the mean, median and mode are all at the same point.

Null hypothesis The starting point of most statistical tests. It assumes that there is no relationship, or no difference, between the factors or data sets being tested. At the end of the test, critical values are used to decide whether to accept or reject the null hypothesis.

Observed data The actual data collected in a survey or fieldwork study. It may be primary or secondary data.

Parametric test A statistical test which assumes that the data comes from a population that follows a normal distribution.

Range The difference between the highest and lowest values in a data set. It is a simple measure of dispersion or spread of the data.

Sampling In a study it is often not possible to investigate the entire population of interest, so a sample must be obtained from which to collect data. There are different types of sampling: **random**, where samples are selected using random numbers generated from a calculator or table; **systematic**, where items are selected at regular intervals; and **stratified**, where measurements are taken from different subsets or 'layers' (such as age groups) of the population.

Scatter graph A type of graph for displaying the likely relationship between two variables. Data is plotted as a series of scatter points, with the values of the independent variable on the horizontal axis and the corresponding values of the dependent variable on the vertical axis. A line of best fit is usually added to show the general trend.

Significance level The level at which you can be confident that your result did not occur purely by chance. In geography, significance levels of 0.05 and 0.01 are typically used, which means that you can be, respectively, 95% or 99% certain that the results did not occur by chance.

Spearman rank correlation test A statistical test for determining the strength of the relationship between two data sets.

Standard deviation A measure of the dispersion or spread of the values in a data set beyond the mean.

Student's _t_ test A parametric statistical test that is used to compare the means of two data sets and determine whether or not there is a significant difference between them.

_x_-axis The horizontal axis of a graph.

_y_-axis The vertical axis of a graph.

Specification cross-reference

The information in the table is intended as a guide only. You should refer to your specification for full details of the topics you need to know.

Basic mathematical skills – prior knowledge useful in the analysis of data in fieldwork or assessments.

Fieldwork skills – all specifications include compulsory fieldwork in the AS and A-level courses. At A-level this forms a non-examined assessment (NEA) or Independent Investigation which represents 20% of the final A-level award. Students must demonstrate knowledge and understanding of techniques used for analysing field data and information, including the ability to select and apply those techniques.

Specified quantitative skills – skills listed in the AS and A-level course content and specific to the examination boards. (* In places OCR uses the term 'such as' which indicates 'for example'.)

Quantitative skills	Basic mathematical skills	Fieldwork skills	OCR AS level	OCR A-level	AQA AS level	AQA A-level	Edexcel AS level	Edexcel A-level	Eduqas AS level	Eduqas A-level
Understanding data										
Nominal, ordinal and interval data	✓	✓								
Ratios and fractions	✓	✓							✓	✓
Percentages	✓	✓							✓	✓
Densities	✓	✓							✓	✓
Measures of central tendency										
Mean, median and mode	✓	✓	✓	✓	✓	✓	✓	✓	✓	✓
Frequency distributions	✓	✓					✓	✓	✓	✓
Measures of dispersion										
Range		✓	✓	✓	✓	✓	✓	✓	✓	✓
Quartile range		✓	✓	✓	✓	✓	✓	✓	✓	✓
Interquartile range		✓	✓	✓	✓	✓	✓	✓	✓	✓
Standard deviation		✓	✓	✓	✓	✓	✓	✓		✓
Measures of concentration										
Nearest neighbour index		✓								
Location quotient		✓								✓
Lorenz curves		✓					✓	✓	✓	✓
Measures of correlation										
Scatter graphs and lines of best fit		✓	✓	✓	✓	✓	✓	✓	✓	✓
Spearman rank correlation coefficient		✓		✓*	✓	✓	✓	✓	✓	✓
Testing for differences between data sets										
Student's t test		✓		✓*			✓	✓		
Mann–Whitney U test		✓		✓*						
Chi-squared test		✓		✓*		✓	✓	✓	✓	✓